P9-BZX-945

Soils and Foundations

Soils
and Foundations

Cheng Liu
and
Jack B. Evett

University of North Carolina at Charlotte

Prentice-Hall, Inc., *Englewood Cliffs, New Jersey 07632*

Library of Congre s Cataloging in Publication Data

Liu, Cheng, 1937–
   Soils and foundations.

   Bibliography: p.
   Includes index.
   1. Soil mechanics. 2. Foundations. I. Evett,
Jack B., 1942– joint author. II. Title.
TA710.L548   624.1'5   80-36715
ISBN 0-13-822239-8

Editorial/production supervision and interior design: Nancy Moskowitz
Manufacturing buyer: Anthony Caruso
Cover design: Edsal Enterprises
Cover photos courtesy of: Soiltest, Inc. and Koehring Company

Printed in the United States of America

10 9 8 7 6 5 4 3 2 1

Prentice-Hall International, Inc., *London*
Prentice-Hall of Australia Pty. Limited, *Sydney*
Prentice-Hall of Canada, Ltd., *Toronto*
Prentice-Hall of India Private Limited, *New Delhi*
Prentice-Hall of Japan, Inc., *Tokyo*
Prentice-Hall of Southeast Asia Pte. Ltd., *Singapore*
Whitehall Books Limited, *Wellington, New Zealand*

To Kimmie, Jonathan, and Michele Liu
    and
Sallie Evett

Contents

Preface

We have attempted to prepare an introductory, practical textbook for soil mechanics and foundations. It is a basic soils and foundations text that emphasizes practical application more than theory, although the practical application is supported by basic theory. Written in a simple and direct style that should make it very easy to read, understand, and grasp the subject matter, this book contains an abundance of both example problems in each chapter and work problems at the end of each chapter. Additionally, there are ample diagrams, charts, and illustrations throughout to help explain the subject matter better. In summary, we have tried to extract the salient and essential aspects of soils and foundations and to present these in a simple and straightforward manner.

We believe this book will be well suited for use by students in both civil engineering and various engineering technology programs. The latter would include both two-year and four-year programs in civil technology, civil engineering technology, and building and/or construction technology. Additionally, we believe it would be useful for some students in other areas such as architecture, geology, geography, agriculture, and so on. In nonacademic contexts, we believe it would be useful for persons such as contractors, perhaps lawyers, architects, and others who deal with soils and foundations and/or civil and soils engineers and technologists and who would like to know more about the subject.

In Chap. 1 we have introduced engineering properties of soils, and in Chap. 2 we have introduced soil exploration. The next three chapters cover stress distribution in soil, consolidation of soil and settlement of structures, and shear strength of soil. These set the stage for the study of foundations, including both shallow foundations (Chap. 6) and deep foundations (pile foundations in Chap. 7 and drilled caissons in Chap. 8). Chapter 9 deals with earth pressure, which paves the way for the study of retaining walls in Chap. 10. The last two chapters cover soil compaction and stability analysis of slopes.

To the student using this book, we urge you to review each illustration as it is cited and especially to study very carefully each example problem. Believing that example problems are an extremely effective means of learning a subject such as soils and foundations, we have included an abundance of example problems, and we believe they will be very useful to you in mastering the material in this book.

We wish to express our sincere appreciation to Carlos G. Bell of the University of North Carolina at Charlotte and to W. Kenneth Humphries of the University of South Carolina, who read our manuscript and offered many helpful suggestions. Also, we thank Judy Craig, who typed the entire manuscript.

We hope you will enjoy using this book. We would be pleased to receive your comments, suggestions, and/or criticisms.

Cheng Liu
Jack B. Evett
Charlotte, North Carolina

Soils and Foundations

1

Engineering Properties of Soils

Soil is more or less taken for granted by the average person. It makes up the ground on which we live, and it makes us dirty. Beyond these observations, most people are not overly concerned with soil. There are, however, certain people who **are** deeply concerned. These include certain engineers and engineering technologists as well as geologists, contractors, hydrologists, and others.

Most structures of all types rest either directly or indirectly upon the soil, and proper analysis of the soil and design of the foundation of the structure are necessary to ensure a safe structure free of undue settling and/or collapse. A comprehensive knowledge of the soil in a specific location is also important in many other contexts. Thus the study of the properties of soils should be an important component in the education of both the civil engineer and the civil engineering technologist.

Chapter 1 is introductory in nature and introduces and defines various engineering properties of soils. Subsequent chapters deal with the evaluation of these properties and with the interrelationships of soil with structures of various types.

1-1 SOIL AND SOIL TYPES

Soil is comprised of particles, large or small, and it may be necessary to include as "soil" not only solid matter but air and water. Normally, these particles are the result of the weathering (disintegration) of rocks and of the decay of vegetation. Some soil particles may, over a period of time, become

1

consolidated under the weight of overlying material and become rock. Even a cycle of rock disintegrating to form soil, soil being consolidated to form rock, rock disintegrating to form soil, and so on, may occur in nature over a geologic period of time. The differentiation between soil and rock is not sharp; but, as a general rule, if material can be removed without blasting, it might be considered as "soil"; whereas, if blasting is required, it might be considered as "rock."

Soils may be separated into three very broad types or categories: *cohesionless* soils, *cohesive* soils, and *organic* soils. In the case of cohesionless soils, the soil particles do not tend to stick together. Cohesive soils are characterized by very small particle size where surface chemical effects predominate. The particles do tend to stick together—the result of water–particle interaction and attractive forces between particles. Cohesive soils are therefore both sticky and plastic. Organic soils are typically spongy, crumbly, and compressible. They are undesirable for use in supporting structures.

Three common types of cohesionless soil are *gravel, sand,* and *silt.* Gravel has particle sizes greater than about 5 millimeters (mm); whereas particle sizes for sand range from about 0.1 mm to 5 mm. Both gravel and sand may be further divided into "fine" (as fine sand) and "coarse" (as coarse sand). Gravel and sand can be classified according to particle size by sieve analysis. Silt has particle sizes that range from about 0.005 mm to 0.1 mm.

The common type of cohesive soil is clay, which has particle sizes less than about 0.005 mm. Clay soils cannot be separated by sieve analysis into size categories because no practical sieve can be made with openings so small; instead, particle sizes may be determined by observing settling velocities of the particles in a water mixture.

Soils can also be categorized in terms of grain size. Two such categories are *coarse-grained* and *fine-grained.* Gravel and sand are coarse-grained, and silt and clay are fine-grained.

In most applications in this book, soils are categorized as either cohesionless or cohesive, with cohesionless generally implying a sandy soil and cohesive, a clayey soil.

1-2 GRAIN-SIZE ANALYSIS
AND ATTERBERG LIMITS

Never will a natural soil be encountered in which all particles are exactly the same size and/or shape. Both cohesionless and cohesive soils will always contain particles of varying sizes. Soils may be encountered that contain both sand and clay, for example. In many engineering applications, it is not sufficient to know only that a given soil is clay, sand, rock, gravel, or silt. It is often necessary to know something about the distribution of grain size of a given soil.

In the case of most cohesionless soils, distribution of grain size can be determined by sieve analysis. A sieve is similar to a cook's flour sifter. A sifter is an apparatus containing a wire mesh with openings the same size and shape. When soil is passed through a sieve, soil particles smaller than the opening size of the sieve will pass through while those larger than the opening size will be retained.

In practice, sieves of varying opening sizes are stacked, with the largest opening size at the top and a pan at the bottom. Soil is poured in at the top, and soil particles pass downward through the sieves until they are retained on a particular sieve. The stack of sieves is mechanically agitated during this procedure. At the end of the procedure, the soil particles retained on each sieve can be weighed and the results presented graphically in the form of a grain-size distribution curve. This is normally a semilog plot with grain size (diameter) along the abscissa on a logarithmic scale and percentage passing that grain size along the ordinate on an arithmetic scale. Example 1-1 illustrates the analysis of the results of a sieve test, including the preparation of a grain-size distribution curve.

EXAMPLE 1-1

Given

An air-dry soil sample weighing 2000 grams (g) is brought to the soils laboratory for mechanical grain-size analysis. The laboratory data are as follows:

U.S. Sieve Size	Size Opening (mm)	Weight Retained (g)
$\frac{3}{4}$ in.	19.1	0
$\frac{3}{8}$ in.	9.52	158
No. 4	4.76	308
No. 10	2.00	608
No. 40	0.425	652
No. 100	0.149	224
No. 200	0.075	42
Pan	—	8

Required

Draw a grain-size distribution curve for the soil sample using five-cycle semilog paper.

Solution

To plot the gradation curve, percentage retained on each sieve, cumulative percentage retained, and percentage passing through each sieve must be calculated and the results tabulated as shown in Table 1-1.

TABLE 1-1   Sieve analysis data.

(1) Sieve Number	(2) Sieve Opening (mm)	(3) Weight Retained (g)	(4) Percentage Retained	(5) Cumulative Percentage Retained	(6) Percentage Passing
$\frac{3}{4}$ in.	19.1	0	0	0	100
$\frac{3}{8}$ in.	9.52	158	7.9	7.9	92.1
No. 4	4.76	308	15.4	23.3	76.7
No. 10	2.00	608	30.4	53.7	46.3
No. 40	0.425	652	32.6	86.3	13.7
No. 100	0.149	224	11.2	97.5	2.5
No. 200	0.075	42	2.1	99.6	0.4
Pan	—	8	0.4	100.0	—

Total sample weight = 2000 g

Note:

(a) The percentage retained on each sieve is obtained by dividing the weight retained on each sieve by the total sample weight. Thus,

$$\text{Percentage retained on } \tfrac{3}{4} \text{ in. sieve} = \frac{0}{2000} \times 100\% = 0\%$$

$$\text{Percentage retained on } \tfrac{3}{8} \text{ in. sieve} = \frac{158}{2000} \times 100\% = 7.9\%$$

$$\text{Percentage retained on No. 4 sieve} = \frac{308}{2000} \times 100\% = 15.4\% \text{ etc.}$$

$$\text{Therefore, column (4)} = \frac{\text{column (3)}}{\text{total sample weight}} \times 100\%.$$

(b) Cumulative percentage retained on each sieve is obtained by summing percentage retained on all coarser sieves. Thus,

Cumulative percentage retained on $\frac{3}{4}$ in. sieve = 0%

Cumulative percentage retained on $\frac{3}{8}$ in. sieve = $0 + 7.9 = 7.9\%$

Cumulative percentage retained on No. 4 sieve = $7.9 + 15.4$
$$= 23.3\%$$

Cumulative percentage retained on No. 10 sieve $= 23.3 + 30.4$
$$= 53.7\% \quad \text{etc.}$$

(c) Percentage passing through each sieve is obtained by subtracting from 100% the cumulative percentage retained on the sieves. Thus,

Percentage passing through $\frac{3}{4}$ in. sieve $= 100\% - 0 = 100\%$

Percentage passing through $\frac{3}{8}$ in. sieve $= 100\% - 7.9 = 92.1\%$

Percentage passing through No. 4 sieve $= 100\% - 23.3 = 76.7\%$
$$\text{etc.}$$

Therefore, column (6) $= 100\% - $ column (5).

(d) Upon completion of these calculations, the grain-size distribution curve is obtained by plotting column (2), sieve opening (mm), versus column (6), percentage passing through, on semilog paper. Percentage passing is always plotted as the ordinate on arithmetic scale and sieve opening as the abscissa on log scale (see Fig. 1-1).

In the case of cohesive soils, distribution of grain size is not determined by sieve analysis because the particles are too small. Particle sizes may be determined by the hydrometer method, which is a process for indirectly observing the settling velocities of the particles in a soil–water mixture. Another valuable technique in analyzing cohesive soils is by use of "Atterberg limits," which will be described in the remainder of this section.

Atterberg [1, 2][1] established four states of consistency for cohesive soils. (Consistency refers to their degree of firmness.) These states are the *liquid*, the *plastic*, the *semisolid*, and the *solid* (Fig. 1-2). The dividing line between the liquid and the plastic states is the *liquid limit*; the dividing line between the plastic and the semisolid states is the *plastic limit*; and the dividing line between the semisolid and the solid states is the *shrinkage limit* (see Fig. 1-2). If a soil in the liquid state is gradually dried out, it will pass through the liquid limit, the plastic state, the plastic limit, the semisolid state, and the shrinkage limit, and will reach the solid state. The liquid, plastic, and shrinkage limits are quantified therefore in terms of water content. For example, the liquid limit is reported in terms of the water content at which the soil changes from the liquid state to the plastic state. The difference between the liquid limit and the plastic limit is the *plasticity index*.

Standard laboratory test procedures are available to determine the liquid, plastic, and shrinkage limits, from which the plasticity index may be determined. These limits and this index are useful numbers in classifying soils and in making judgments for their applications.

---

[1] Numbers in brackets refer to the references listed at the end of each chapter.

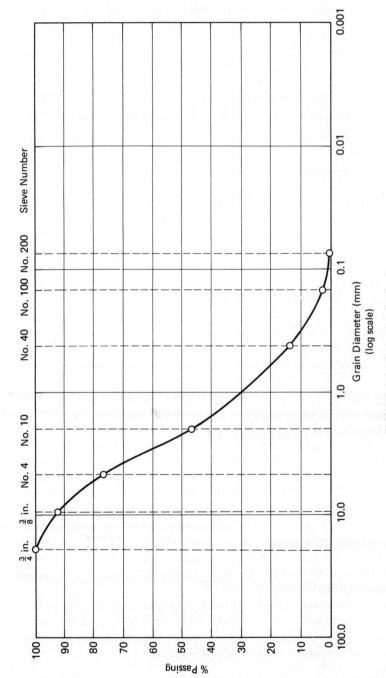

FIGURE 1-1 Grain-size distribution curve.

6

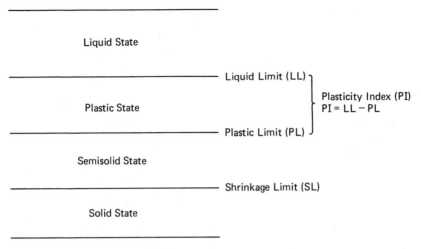

FIGURE 1-2   Atterburg limits. [2]

1-3 SOIL CLASSIFICATION SYSTEMS

In order to be able to describe, in general, a specific soil without listing values of the many soil parameters, it would be convenient to have some kind of generalized "classification system." In practice, there have developed a number of different classification systems, most of which were developed to meet the specific needs of the particular group that developed the system. Some of these soil classification systems are presented briefly in this section.

AASHTO Classification System [3]

AASHTO stands for the American Association of State Highway and Transportation Officials, and therefore this classification system is widely used in highway work. Required parameters for classification by this system are grain size analysis, liquid limit, and plasticity index. With values of these parameters known, one enters the first (left) column of Table 1-2 and determines whether or not the known parameters meet the limiting values in that column. If they do, then the soil classification is that given at the top of the column (A-1-a, if the known parameters meet the limiting values in the first column). If they do not, one enters the next column (to the right) and determines whether or not the known parameters meet the limiting values in that column. The procedure is repeated until the **first** column is reached in which the known parameters meet the limiting values in that column. The particular soil classification is that given at the top of that particular column.

TABLE 1-2 Classification of soils and soil–aggregate mixtures [3].

General Classification	Granular Materials (35% or less passing 0.075 mm)							Silt–Clay Materials (more than 35% passing 0.075 mm)			
	A-1		A-3	A-2				A-4	A-5	A-6	A-7
Group Classification	A-1-a	A-1-b		A-2-4	A-2-5	A-2-6	A-2-7				A-7-5, A-7-6
Sieve analysis: percent passing:											
2.00 mm (No. 10)	50 max.	—	—								
0.425 mm (No. 40)	30 max.	50 max.	51 min.	—	—	—	—				
0.075 mm (No. 200)	15 max.	25 max.	10 max.	35 max.	35 max.	35 max.	35 max.	36 min.	36 min.	36 min.	36 min.
Characteristics of fraction passing 0.425 mm (No. 40)											
Liquid limit			—	40 max.	41 min.	40 max.	41 min.	40 max.	41 min.	40 max.	41 min.
Plasticity index	6 max.		N.P.	10 max.	10 max.	11 min.	11 min.	10 max.	10 max.	11 min.	11 min.[1]
Usual types of significant constituent materials	Stone fragments, gravel, and sand		Fine sand	Silty or clayey gravel and sand				Silty soils		Clayey soils	
General rating as subgrade	Excellent to good							Fair to poor			

[1]Plasticity index of A-7-5 subgroup is equal to or less than LL minus 30. Plasticity index of A-7-6 subgroup is greater than LL minus 30.

Once a soil has been classified using Table 1-2, it can be further described using the "group index." The group index utilizes the percent of soil passing a No. 200 sieve, the liquid limit, and the plasticity index. Using known values of these parameters, the group index is computed from the equation

$$\text{group index} = (F - 35)[0.2 + 0.005(LL - 40)] \\ + 0.01(F - 15)(PI - 10) \tag{1-1}$$

where $F =$ percentage of soil passing a No. 200 sieve

$LL =$ liquid limit

$PI =$ plasticity index

The group index computed from Eq. (1-1) is rounded off to the nearest whole number and is appended in parentheses to the group designation determined from Table 1-2. If the computed group index is either zero or negative, the number zero is used as the group index and should be appended to the group designation. If preferred, Fig. 1-3 may be used instead of Eq. (1-1) to determine the group index.

As a general rule, the value of soil as a subgrade material is in inverse ratio to its group index.

Unified Soil Classification System [4, 5, 6]

This system was developed by Casagrande and is utilized by the Corps of Engineers. It utilizes letter symbols to classify soils. The letter symbols are as follows:

G   gravel

S   sand

M   silt

C   clay

W   well graded

P   poorly graded

U   uniformly graded

L   low liquid limit

H   high liquid limit

Normally, two letter symbols are used to classify a soil. For example, SW indicates well graded sand. In classifying silts and clays, the liquid limit and

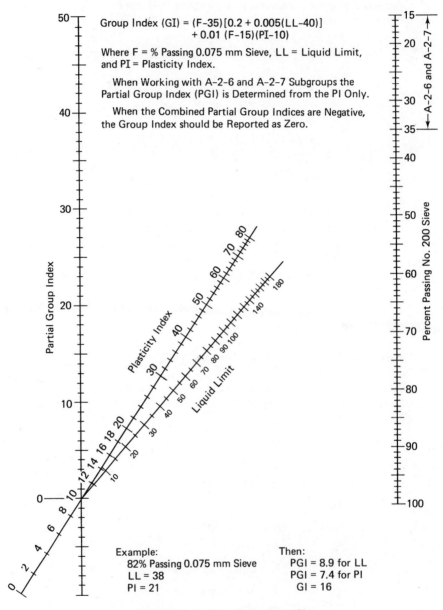

Group Index (GI) = (F–35)[0.2 + 0.005(LL–40)] + 0.01 (F–15)(PI–10)

Where F = % Passing 0.075 mm Sieve, LL = Liquid Limit, and PI = Plasticity Index.

When Working with A-2-6 and A-2-7 Subgroups the Partial Group Index (PGI) is Determined from the PI Only.

When the Combined Partial Group Indices are Negative, the Group Index should be Reported as Zero.

Example:
82% Passing 0.075 mm Sieve
LL = 38
PI = 21

Then:
PGI = 8.9 for LL
PGI = 7.4 for PI
GI = 16

FIGURE 1-3 Group index chart. [3]

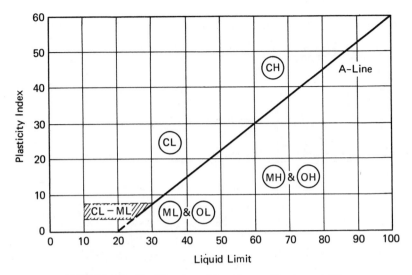

FIGURE 1-4 Plasticity chart. (From Corps of Engineers.) [6]

plastic limit are used in conjunction with the plasticity chart shown in Fig. 1-4. A chart for laboratory identification procedure is given in Fig. 1-5.

Example 1-2 illustrates the AASHTO classification system and the Unified soil classification system.

EXAMPLE 1-2

Given

A sample of soil was tested in the laboratory and the results of the laboratory tests were as follows:

1. Liquid limit = 42.3%.
2. Plastic limit = 15.8%.
3. The following sieve analysis data:

U.S. Sieve Size	Percentage Passing
No. 4	100
No. 10	93.2
No. 40	81.0
No. 200	60.2

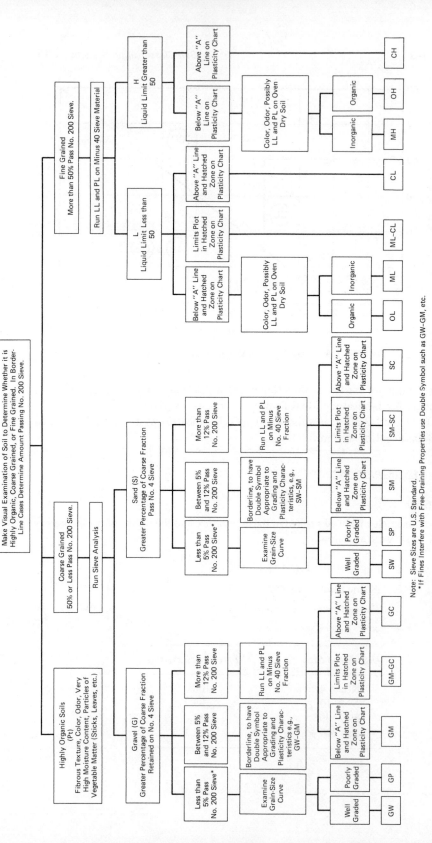

FIGURE 1-5 Chart for auxiliary laboratory identification procedure. (From Corps of Engineers.) [6]

Note: Sieve Sizes are U.S. Standard.
*If Fines Interfere with Free-Draining Properties use Double Symbol such as GW-GM, etc.

Required

Classify the soil sample by

1. The AASHTO classification system.
2. The Unified soil classification system.

Solution

1. *By the AASHTO classification system:*
   From Table 1-2, the sample is classified as A-7. According to the AASHTO classification system, the plasticity index of the A-7-5 subgroup is equal to or less than the liquid limit minus 30, and the plasticity index of the A-7-6 subgroup is greater than the liquid limit minus 30 (see footnote under Table 1-2).

   plasticity index (PI) = liquid limit (LL) − plastic limit (PL)

   $PI = 42.3 - 15.8 = 26.5\%$

   $LL - 30 = 42.3 - 30 = 12.3\%$

   $[PI = 26.5\%] > [LL - 30 = 12.3\%]$

   Hence, this is A-7-6 material.
   From Fig. 1-3 (group index chart), with $LL = 42.3\%$ and percentage passing No. 200 sieve = 60.2%, partial group index for $LL = 5.3$. With $PI = 26.5\%$ and percentage passing No. 200 sieve = 60.2%, partial group index for $PI = 7.5$.

   $$\text{Total group index} = 5.3 + 7.5 = 12.8$$

   Hence, the soil is A-7-6 (13).

2. *By the Unified soil classification system:*
   Since the percentage passing the No. 200 sieve is 60.2%, which is greater than 50%, go to the first block (labeled "Fine-grained") in the right column of Fig. 1-5. Now, since the liquid limit is 42.3%, which is less than 50%, go downward in Fig. 1-5 to the block labeled "L." Referring next to the plasticity chart (Fig. 1-4), the sample is located above A-line and the hatched zone. Returning to Fig. 1-5, go downward to the block labeled "CL." Thus, the soil is classified, CL, according to the Unified soil classification system.

Federal Aviation Administration (FAA) Classification [6, 7]

This classification of soils is utilized by the FAA in airport construction. In this system, soil groups are designated as E-1, E-2, E-3, . . . , E-13, as determined from Table 1-3 and Fig. 1-6. As was the case with the aforementioned soil classification systems, the soil parameters of grain-size analysis,

TABLE 1-3   Classification of soils for airport construction[1] [6].

MECHANICAL ANALYSIS

MATERIAL FINER THAN No. 10 SIEVE

Soil Group	Retained on No. 10 Sieve[2] (%)	Coarse Sand, Pass No. 10 Retained on No. 40 (%)	Fine Sand, Pass No. 40 Retained on No. 200 (%)	Combined Silt and Clay, Pass No. 200 (%)	Liquid Limit	Plasticity Index
E-1	0–45	40+	60–	15–	25–	6–
E-2	0–45	15+	85–	25–	25–	6–
E-3	0–45	—	—	25–	25–	6–
E-4	0–45	—	—	35–	35–	10–
E-5	0–55	—	—	45–	40–	15–
E-6	0–55	—	—	45+	40–	10–
E-7	0–55	—	—	45+	50–	10–30
E-8	0–55	—	—	45+	60–	15–40
E-9	0–55	—	—	45+	40+	30–
E-10	0–55	—	—	45+	70–	20–50
E-11	0–55	—	—	45+	80–	30+
E-12	0–55	—	—	45+	80+	—
E-13	Muck and peat—field examination					

[1]Courtesy Federal Aviation Administration.
[2]Classification is based on sieve analysis of the portion of the sample passing the No. 10 sieve. When a sample contains material coarser than the No. 10 sieve in amounts equal to or greater than the maximum limit shown in the table, a raise in classification may be allowed provided the coarse material is reasonably sound and fairly well graded.

liquid and plastic limits, and plasticity index are used for the FAA soil classification system. The use of Table 1-3 and Fig. 1-6 should be more or less self-explanatory.

Although there are other soil classification systems available, the three presented here are commonly used. It should be noted that such soil classification systems are quite general in nature. For example, if a soil is classified as SW (Unified soil classification system), a well graded sand is indicated. However, various tests would have to be performed to determine specific (quantitative) characteristics of this soil. Nevertheless, soil classification systems are important and are routinely used to estimate behavior of soil.

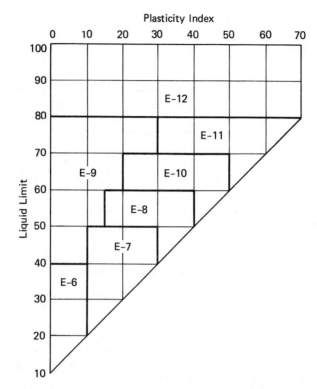

FIGURE 1-6 FAA classification chart for fine-grained soils. (Courtesy Federal Aviation Administration.) [6]

1-4 COMPONENTS OF SOILS

Soils contain three components, which may be characterized as solid, liquid, and gas. As indicated in Sec. 1-1, the solid components of soil are weathered rock and (sometimes) decayed vegetation. The liquid component of soils is almost always water (often with dissolved matter), and the gas component is air. The volume of water and air combined is referred to as the *void.*

Figure 1-7 gives a block diagram showing the components of soil. These components may be considered in terms of their volumes and in terms of their weights. In Fig. 1-7, the terms V, V_a, V_w, V_s, and V_v represent total volume, volume of air, volume of water, volume of solid matter and volume of voids, respectively. The terms W, W_a, W_w, and W_s represent total weight, weight of air, weight of water, and weight of solid matter, respectively. The weight of air (W_a) is virtually zero.

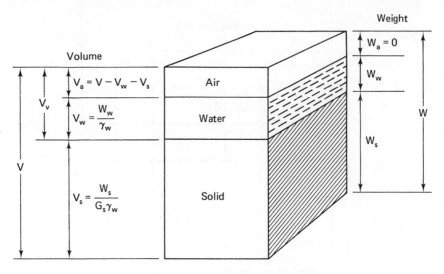

FIGURE 1-7 Block diagram showing components of soil.

1-5 WEIGHT–VOLUME RELATIONSHIPS

A number of important relationships exist among the components of soil in terms of both weight and volume. These relationships define new parameters that are useful in working with soils.

In terms of volume, the following new parameters are important—the *void ratio*, *porosity*, and *degree of saturation*. The void ratio (e) is the ratio (expressed as a decimal fraction) of the volume of voids to the volume of solids.

$$e = \frac{V_v}{V_s} \qquad (1\text{-}2)$$

Porosity (n) is the ratio (expressed as a percentage) of the volume of voids to the total volume.

$$n = \frac{V_v}{V} \times 100 \qquad (1\text{-}3)$$

The degree of saturation (S) is the ratio (expressed as a percentage) of the volume of water to the volume of voids.

$$S = \frac{V_w}{V_v} \times 100 \qquad (1\text{-}4)$$

In terms of weight, the following new parameters are important—*water content, unit weight* (also referred to as *density*), *dry unit weight* (also referred to as *dry density*), and *specific gravity of the solids*. (*Note:* The term "unit weight," or "density," implies "**wet** unit weight," or "**wet** density." If "dry unit weight," or "dry density," is intended, the adjective "dry" is indicated explicitly.) The water content (w) is the ratio (expressed as a percentage) of the weight of water to the weight of solids.

$$w = \frac{W_w}{W_s} \times 100 \qquad (1\text{-}5)$$

The unit weight (γ) is the total weight (weight of solid plus weight of water) divided by the total volume (volume of solid plus volume of water plus volume of air).

$$\gamma = \frac{W}{V} \qquad (1\text{-}6)$$

The dry unit weight (γ_d) is the weight of solids divided by the total volume.

$$\gamma_d = \frac{W_s}{V} \qquad (1\text{-}7)$$

The specific gravity of the solids (G_s) is the ratio of the unit weight of the solids (weight of solids divided by volume of solids) to the unit weight of water.

$$G_s = \frac{W_s/V_s}{\gamma_w} = \frac{W_s}{V_s \gamma_w} \qquad (1\text{-}8)$$

where γ_w is the unit weight of water [62.4 pounds per cubic foot (lb/ft³, or pcf) or 1 gram per cubic centimeter (g/cm³)].

The soils engineer or technologist must be proficient in determining these parameters based on laboratory evaluations of weight and volume of the components of a soil. The use of a block diagram (as shown in Fig. 1-7) is recommended to help obtain answers more quickly and more accurately. Eight example problems follow.

EXAMPLE 1-3

Given

1. The weight of a chunk of moist soil sample is 45.6 lb.
2. The volume of the soil chunk measured before drying is 0.40 ft³.
3. After being dried out in an oven, the weight of dry soil is 37.8 lb.
4. The specific gravity of the solids is 2.65.

Required

1. The water content.
2. The unit weight of moist soil.
3. The void ratio.
4. The porosity.
5. The degree of saturation.

Solution

See Fig. 1-8. (Boldface data on the figure indicate given information. Other data are calculated in the solution of the problem.)

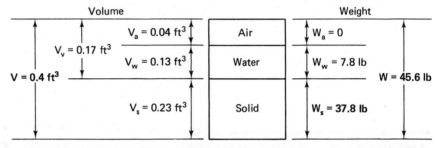

FIGURE 1-8

1. Water content $(w) = \dfrac{W_w}{W_s} \times 100 = \dfrac{45.6 - 37.8}{37.8} \times 100 = 20.6\%$

2. Unit weight of moist soil $(\gamma) = \dfrac{W}{V} = \dfrac{45.6}{0.40} = 114.0 \text{ lb/ft}^3$

3. $V_w = \dfrac{W_w}{\gamma_w} = \dfrac{45.6 - 37.8}{62.4} = 0.13 \text{ ft}^3$

   $V_s = \dfrac{W_s}{G_s \gamma_w} = \dfrac{37.8}{(2.65)(62.4)} = 0.23 \text{ ft}^3$

   $V_a = V - V_w - V_s = 0.40 - 0.13 - 0.23 = 0.04 \text{ ft}^3$

   $V_v = V - V_s = 0.40 - 0.23 = 0.17 \text{ ft}^3$

   or

   $V_v = V_a + V_w = 0.04 + 0.13 = 0.17 \text{ ft}^3$

   Void ratio $(e) = \dfrac{V_v}{V_s} = \dfrac{0.17}{0.23} = 0.74$

4. Porosity $(n) = \dfrac{V_v}{V} \times 100 = \dfrac{0.17}{0.40} \times 100 = 42.5\%$

5. Degree of saturation $(S) = \dfrac{V_w}{V_v} \times 100 = \dfrac{0.13}{0.17} \times 100 = 76.5\%$

EXAMPLE 1-4

Given

1. The wet weight of a soil specimen is 207 g.
2. The volume of the specimen measured before drying is 110 cm³.
3. The dried weight of the specimen is 163 g.
4. The specific gravity of the solids is 2.68.

Required

1. The void ratio.
2. The degree of saturation.
3. The wet density in pcf.
4. The dry density in pcf.

Solution

See Fig. 1-9.

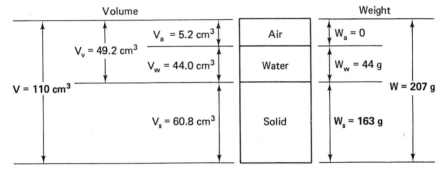

FIGURE 1-9

1. $V_s = \dfrac{W_s}{G_s\gamma_w} = \dfrac{163}{(2.68)(1)} = 60.8$ cm³

$$V_w = \dfrac{W_w}{\gamma_w} = \dfrac{207 - 163}{1} = 44.0 \text{ cm}^3$$

$$V_a = V - V_w - V_s = 110 - 44.0 - 60.8 = 5.2 \text{ cm}^3$$

$$V_v = V - V_s = 110 - 60.8 = 49.2 \text{ cm}^3$$

or

$$V_v = V_a + V_w = 5.2 + 44.0 = 49.2 \text{ cm}^3$$

$$\text{Void ratio } (e) = \dfrac{V_v}{V_s} = \dfrac{49.2}{60.8} = 0.81$$

2. Degree of saturation $(S) = \dfrac{V_w}{V_v} \times 100 = \dfrac{44.0}{49.2} \times 100 = 89.4\%$

3. Wet density $(\gamma_{wet}) = \dfrac{W}{V} = \dfrac{207}{110} = 1.88$ g/cm³

   wet density $(\gamma_{wet}) = 1.88 \times 62.4 = 117.3$ pcf

4. Dry density $(\gamma_d) = \dfrac{W_s}{V} = \dfrac{163}{110} = 1.48$ g/cm³

   dry density $(\gamma_d) = 1.48 \times 62.4 = 92.4$ pcf

EXAMPLE 1-5

Given

1. A soil specimen has a water content of 12% and a wet unit weight (wet density) of 127 pcf.
2. The specific gravity of the solids of the soil specimen is 2.69.

Required

1. The dry density (γ_d).
2. The void ratio (e).
3. The degree of saturation (S).

Solution

See Fig. 1-10.

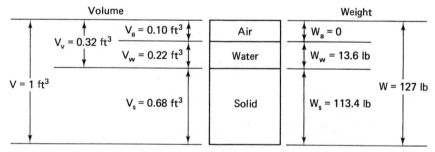

FIGURE 1-10

Water content $(w) = \dfrac{W_w}{W_s} \times 100$

$$12 = \dfrac{W_w}{W_s} \times 100$$

$$\therefore \dfrac{W_w}{W_s} = 0.12 \qquad\qquad \text{(A)}$$

$$W = W_w + W_s$$
$$127 = W_w + W_s \quad \text{(assuming a volume of 1 ft}^3\text{)} \tag{B}$$

Solve simultaneous equations (A) and (B). From (A), $W_w = 0.12W_s$. Substituting into (B) gives

$$127 = 0.12W_s + W_s = 1.12W_s$$
$$W_s = 113.4 \text{ lb}$$
$$W_w = 0.12W_s = (0.12)(113.4) = 13.6 \text{ lb}$$

1. Dry density $(\gamma_d) = \dfrac{W_s}{V} = \dfrac{113.4}{1} = 113.4 \text{ pcf}$

2. $V_s = \dfrac{W_s}{G_s\gamma_w} = \dfrac{113.4}{(2.69)(62.4)} = 0.68 \text{ ft}^3$

   $V_w = \dfrac{W_w}{\gamma_w} = \dfrac{13.6}{62.4} = 0.22 \text{ ft}^3$

   $V_a = V - V_w - V_s = 1 - 0.22 - 0.68 = 0.10 \text{ ft}^3$

   $V_v = V - V_s = 1 - 0.68 = 0.32 \text{ ft}^3$

   or

   $V_v = V_a + V_w = 0.10 + 0.22 = 0.32 \text{ ft}^3$

   Void ratio $(e) = \dfrac{V_v}{V_s} = \dfrac{0.32}{0.68} = 0.47$

3. Degree of saturation $(S) = \dfrac{V_w}{V_v} \times 100 = \dfrac{0.22}{0.32} \times 100 = 68.8\%$

EXAMPLE 1-6

Given

An undisturbed soil sample has the following data:

1. The void ratio = 0.78.
2. The water content = 12%.
3. The specific gravity of the solids = 2.68.

Required

1. The wet unit weight in pcf.
2. The dry unit weight in pcf.
3. The degree of saturation.
4. The porosity.

Solution

See Fig. 1-11.

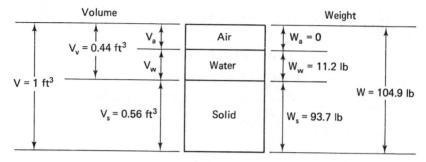

FIGURE 1-11

Since the void ratio $(e) = 0.78$,

$$\frac{V_v}{V_s} = 0.78; \qquad V_v = 0.78V_s \qquad\qquad (A)$$

$$V_v + V_s = V = 1 \text{ ft}^3 \qquad\qquad (B)$$

Substitute Eq. (A) into Eq. (B).

$$0.78V_s + V_s = 1$$

$$V_s = \frac{1}{1.78} = 0.56 \text{ ft}^3$$

$$V_v = 1 - 0.56 = 0.44 \text{ ft}^3$$

$$V_s = \frac{W_s}{G_s\gamma_w}; \qquad 0.56 = \frac{W_s}{(2.68)(62.4)}$$

$$W_s = (0.56)(2.68)(62.4) = 93.7 \text{ lb}$$

From given water content, $\dfrac{W_w}{W_s} = 0.12$,

$$W_w = 0.12W_s = (0.12)(93.7) = 11.2 \text{ lb}$$

1.   Wet unit weight $(\gamma_{\text{wet}}) = \dfrac{W}{V} = \dfrac{W_w + W_s}{V} = \dfrac{11.2 + 93.7}{1} = 104.9 \text{ pcf}$

2.   Dry unit weight $(\gamma_d) = \dfrac{W_s}{V} = \dfrac{93.7}{1} = 93.7 \text{ pcf}$

3. $V_w = \dfrac{W_w}{\gamma_w} = \dfrac{11.2}{62.4} = 0.18 \text{ ft}^3$

Degree of saturation $(S) = \dfrac{V_w}{V_v} \times 100 = \dfrac{0.18}{0.44} \times 100 = 40.9\%$

4.   Porosity $(n) = \dfrac{V_v}{V} \times 100 = \dfrac{0.44}{1} \times 100 = 44.0\%$

EXAMPLE 1-7

Given

    1.   A 100% saturated soil has a wet unit weight of 120 pcf.

    2.   The water content of this saturated soil was determined to be 36%.

Required

    1.   Void ratio.

    2.   The specific gravity of the solids.

Solution

    See Fig. 1-12.

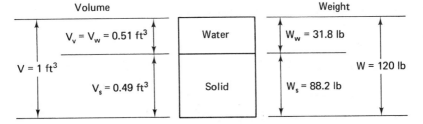

FIGURE 1-12

$$W_w + W_s = 120 \text{ lb} \qquad\qquad\qquad \text{(A)}$$

$$\frac{W_w}{W_s} = 0.36 \qquad\qquad\qquad \text{(B)}$$

From (B), $W_w = 0.36 W_s$; substitute into (A).

$$0.36 W_s + W_s = 120 \text{ lb}$$

$$W_s = \frac{120}{1.36} = 88.2 \text{ lb}$$

$$W_w = 0.36 W_s = (0.36)(88.2) = 31.8 \text{ lb}$$

1. $V_w = \dfrac{W_w}{\gamma_w} = \dfrac{31.8}{62.4} = 0.51 \text{ ft}^3$

$$V_s = V - V_w = 1 - 0.51 = 0.49 \text{ ft}^3$$

$$e = \frac{V_v}{V_s} = \frac{V_w}{V_s} = \frac{0.51}{0.49} = 1.04$$

Note: In this problem, because the soil is 100% saturated, $V_v = V_w$.

2. $V_s = \dfrac{W_s}{(G_s)(\gamma_w)}$; $0.49 = \dfrac{88.2}{(G_s)(62.4)}$

$G_s = \dfrac{88.2}{(0.49)(62.4)} = 2.88$

EXAMPLE 1-8

Given

    1. A 100% saturated soil.

    2. Water content = 34%.

    3. Void ratio = 0.92.

Required

    1. Unit weight of soil.

    2. The specific gravity of the solids.

Solution

See Fig. 1-13.

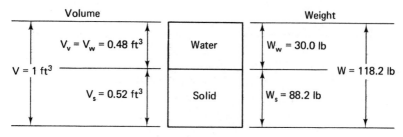

FIGURE 1-13

$$\frac{V_v}{V_s} = \frac{V_w}{V_s} = 0.92 \quad \text{(given)}$$

$$V_v = 0.92 V_s \tag{A}$$

$$V_v + V_s = 1.0 \text{ ft}^3 \tag{B}$$

Substitute Eq. (A) into Eq. (B).

$$0.92 V_s + V_s = 1.0$$

$$V_s = 0.52 \text{ ft}^3$$

$$V_v = V_w = 1 - 0.52 = 0.48 \text{ ft}^3$$

$$W_w = (V_w)(62.4) = (0.48)(62.4) = 30.0 \text{ lb}$$

$$\frac{W_w}{W_s} = 0.34 \quad \text{(given)}$$

$$W_s = \frac{W_w}{0.34} = \frac{30.0}{0.34} = 88.2 \text{ lb}$$

$$W = W_w + W_s = 30.0 + 88.2 = 118.2 \text{ lb}$$

1. Unit weight $= \dfrac{W}{V} = \dfrac{118.2}{1} = 118.2 \text{ pcf}$

2. $V_s = \dfrac{W_s}{(G_s)(\gamma_w)};$ $0.52 = \dfrac{88.2}{(G_s)(62.4)}$

$$G_s = \frac{88.2}{(0.52)(62.4)} = 2.72$$

EXAMPLE 1-9

Given

A 100% saturated soil is characterized by the following data:

1. Water content $= 37.5\%$.
2. The specific gravity of the solids $= 2.72$.

Required

1. Void ratio.
2. Unit weight.

Solution

See Fig. 1-14.

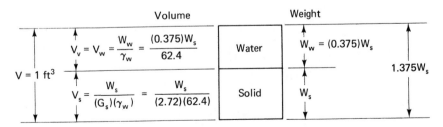

FIGURE 1-14

1. Water content $= \dfrac{W_w}{W_s} \times 100$

$$\dfrac{W_w}{W_s} \times 100 = 37.5 \quad \text{(given)}$$

$$\dfrac{W_w}{W_s} = 0.375 \qquad \therefore \; W_w = 0.375 W_s$$

$$V_v = V_w = \dfrac{W_w}{\gamma_w} = \dfrac{(0.375)(W_s)}{62.4}$$

$$V_s = \dfrac{W_s}{(G_s)(\gamma_w)} = \dfrac{W_s}{(2.72)(62.4)}$$

$$e = \dfrac{V_v}{V_s} = \dfrac{(0.375)(W_s)/62.4}{W_s/(2.72)(62.4)} = (0.375)(2.72) = 1.02$$

2. $e = \dfrac{V_v}{V_s} = 1.02 \qquad \therefore \; V_v = 1.02 V_s$ \hfill (A)

$$V_v + V_s = 1 \text{ ft}^3 \hfill \text{(B)}$$

Substitute (A) into (B).

$$1.02 V_s + V_s = 1$$

$$V_s = 0.495 \text{ ft}^3 \quad \text{and} \quad V_v = V_w = 1 - 0.495 = 0.505 \text{ ft}^3$$

$$W_w = (V_w)(\gamma_w) = (0.505)(62.4) = 31.5 \text{ lb}$$

$$\text{Water content} = \dfrac{W_w}{W_s} \times 100 = 37.5\%$$

$$W_s = \dfrac{W_w}{0.375} = \dfrac{31.5}{0.375} = 84.0 \text{ lb}$$

$$\text{Unit weight} = \dfrac{W}{V} = \dfrac{W_w + W_s}{V} = \dfrac{31.5 + 84.0}{1} = 115.5 \text{ pcf}$$

EXAMPLE 1-10

Given

A soil sample has the following data:

1. Void ratio $= 0.94$.
2. Degree of saturation $= 35\%$.
3. The specific gravity of the solids $= 2.71$.

Required

1. Water content.
2. Unit weight.

Solution

See Fig. 1-15.

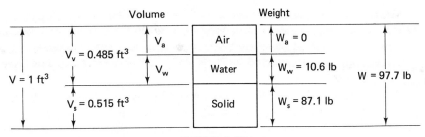

FIGURE 1-15

From given void ratio, $e = \dfrac{V_v}{V_s} = 0.94,$

$$V_v = 0.94 V_s \qquad\qquad \text{(A)}$$
$$V_v + V_s = 1 \text{ ft}^3 \qquad\qquad \text{(B)}$$

Substitute Eq. (A) into (B).

$$0.94 V_s + V_s = 1$$
$$V_s = 0.515 \text{ ft}^3$$
$$V_v = 0.485 \text{ ft}^3$$

From given degree of saturation, $S = \dfrac{V_w}{V_v} = 0.35,$

$$V_w = 0.35 V_v$$
$$V_w = (0.35)(0.485) = 0.170 \text{ ft}^3$$
$$W_w = (V_w)(\gamma_w) = (0.170)(62.4) = 10.6 \text{ lb}$$
$$W_s = (V_s)(G_s)(\gamma_w) = (0.515)(2.71)(62.4) = 87.1 \text{ lb}$$

1. Water content $= \dfrac{W_w}{W_s} \times 100 = \dfrac{10.6}{87.1} \times 100 = 12.2\%$

2. Unit weight $= \dfrac{W}{V} = \dfrac{W_w + W_s}{V} = \dfrac{10.6 + 87.1}{1} = 97.7 \text{ pcf}$

1-6 COMPRESSIBILITY

When soil is compressed, its volume is decreased. The decrease in volume of the soil is the reduction in voids within the soil and consequently can be expressed as a reduction in the void ratio (e). Soil compression, which results from a loading and causes a reduction in the volume of voids (or decrease in void ratio), is usually brought on by the extruding of water and/or air from soil. If a saturated soil is subjected to the weight of a building and water is subsequently squeezed out or otherwise lost, resulting soil compression can cause undue building settlement. If water is added to the soil, soil expansion may occur causing building uplift.

Compressibility is more pronounced in the case of cohesive soils, where soil moisture plays a part in particle interaction. The ultimate volume decrease may not occur until some time after loading. In the case of cohesionless soils, compressibility is less, and ultimate volume decrease occurs at or immediately after loading.

The preceding discussion of compressibility of soil is presented here to give a brief introduction to this subject, since it is the purpose of this chapter to introduce the various engineering properties of soils. A more comprehensive treatment of compressibility is given in Chap. 4.

1-7 SHEAR STRENGTH [8][1]

Shear strength of a soil refers to its ability to resist shear stresses. Shear stresses exist in a sloping hillside or result from filled land, weight of footings, and so on. If a given soil does not have sufficient shear strength to resist such shear stresses, failures in the forms of landslides and footing failures will occur.

Soil gains its shear strength from two sources—internal friction and cohesion. (The internal friction includes sliding and rolling friction and the resistance offered by interlocking action among soil particles.) This may be exhibited in equation form by Coulomb's equation:

$$s = c + \sigma' \tan \phi \qquad (1\text{-}9)$$

where $s = $ shear strength, pounds per square foot (lb/ft², or psf)

$c = $ cohesion, psf

$\sigma' = $ effective intergranular normal (perpendicular to the shear plane) pressure, psf

$\phi = $ angle of internal friction, degrees (deg)

$\tan \phi = $ coefficient of friction

This equation is represented graphically by line A in Fig. 1-16.

In the case of cohesionless soil (such as sand), there is virtually no cohesion ($c = 0$) and Eq. (1-9) reverts to: $s = \sigma' \tan \phi$. This is represented graphically by line B in Fig. 1-16. In the case of cohesive soil (such as clay), the

---

[1] Wayne C. Teng, *Foundation Design*, © 1962. Reprinted by permission of Prentice-Hall, Inc. (This footnote applies to all succeeding citations to this reference in this book.)

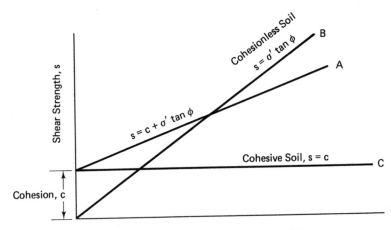

FIGURE 1-16 Shear strength diagram. [8]

angle of internal friction (ϕ) can be taken to be zero for many foundation design problems. If ϕ is taken to be zero, Eq. (1-9) reverts to: $s = c$. This is represented graphically by line C in Fig. 1-16.

The preceding discussion of shear strength of cohesionless soil and of cohesive soil is presented here to give an introduction to the shear strength of soil. A more comprehensive treatment of the shear strength of both cohesionless and cohesive soils, including certain long-term effects on the shear strength of cohesive soil, is given in Chap. 5.

The shear strength parameters, c and ϕ, in Eq. (1-9) can be determined directly or indirectly by standard field or laboratory tests (see Chap. 5).

1-8 PROBLEMS

1-1 Draw a gradation curve (use three-cycle semilog paper) for a soil sample that has the following test data for mechanical grain-size analysis:

U.S. Sieve Size	Size Opening (mm)	Weight Retained (g)
$\frac{3}{8}$ in.	9.52	0
No. 4	4.76	42
No. 10	2.00	146
No. 40	0.420	458
No. 100	0.149	218
No. 200	0.074	73
Pan	—	63

1-2 A sample of soil was tested in the laboratory and the results of the tests were listed as follows. Classify the soil by both the AASHTO and the Unified soil classification systems.

1. Liquid limit $= 29\%$.

2. Plastic limit $= 19\%$.

3. Mechanical grain-size analysis:

U.S. Sieve Size	Percentage Passing
1 in.	100
$\frac{3}{4}$ in.	90
$\frac{3}{8}$ in.	82
No. 4	70
No. 10	65
No. 40	54
No. 200	25

1-3 An undisturbed chunk of soil has a wet weight of 62 lb and a volume of 0.56 ft^3. When dried out in an oven, the soil weighs 50 lb. If the specific gravity of the solids is found to be 2.64, determine the water content, wet unit weight of soil, dry unit weight of soil, void ratio, porosity, and degree of saturation.

1-4 A 72-cm^3 sample of moist soil weighs 141.5 g. When it is dried out in an oven, it weighs 122.7 g. The specific gravity of the solids is found to be 2.66. Find the water content, void ratio, porosity, degree of saturation, and wet and dry densities in pcf.

1-5 A soil specimen has a water content of 18% and a wet unit weight of 118.5 pcf. The specific gravity of the solids is found to be 2.72. Find the dry density (dry unit weight), void ratio, and degree of saturation.

1-6 An undisturbed soil sample has a void ratio of 0.56, water content of 15%, and specific gravity of the solids of 2.64. Find the wet and dry unit weights in pcf, the porosity, and the degree of saturation.

1-7 A 100% saturated soil has a wet unit weight of 112.8 pcf, and its water content is 42%. Find the void ratio and the specific gravity of the solids.

1-8 A 100% saturated soil has a void ratio of 1.33 and a water content of 48%. Find the unit weight of soil and the specific gravity of the solids.

1-9 The water content of a 100% saturated soil is 35% and the specific gravity of the solids is 2.70. Determine the void ratio and the unit weight.

1-10   A soil sample has the following data:

1. Degree of saturation $= 42\%$.
2. Void ratio $= 0.85$.
3. Specific gravity of the solids $= 2.74$.

Find its water content and unit weight.

References

[1]   A. ATTERBERG, Various papers published in the *Int. Mitt. Bodenkd*, 1911, 1912.

[2]   B. K. HOUGH, *Basic Soils Engineering*, 2nd ed., The Ronald Press Company, New York, 1969. Copyright © 1969, John Wiley & Sons, Inc. Reprinted by permission of John Wiley & Sons, Inc.

[3]   *Standard Specifications for Transportation Materials and Methods of Sampling and Testing, Part I, Specifications*, 12th ed., AASHTO, 1978.

[4]   A. CASAGRANDE, "Classification and Identification of Soils," *Trans. ASCE*, **113**, 901 (1948).

[5]   U.S. Army Corps of Engineers, *The Unified Soil Classification System*, Waterways Exp. Sta. Tech. Mem. 3-357, Vicksburg, Miss., 1953.

[6]   E. J. YODER AND M. W. WITCZAK, *Principles of Pavement Design*, 2nd ed., John Wiley & Sons, Inc., New York, 1975. Copyright © 1975, John Wiley & Sons, Inc. Reprinted by permission of John Wiley & Sons, Inc.

[7]   *Airport Paving*, AC 150/5320-6 H, Federal Aviation Administration, U.S. Department of Transportation, 1967.

[8]   WAYNE C. TENG, *Foundation Design*, Prentice-Hall, Inc., Englewood Cliffs, N.J., 1962.

2

Soil Exploration

2-1 INTRODUCTION

In Chap. 1, various engineering properties of soils were presented. An evaluation of these properties is absolutely necessary in any rational design of structures resting on, in, or against soil. In order to evaluate these properties, it is imperative that the soils engineer or technologist visit the proposed construction site and collect and test soil samples, in order to evaluate and record the results in a useful and meaningful form.

Chapter 2 deals with this evaluation of soil properties, including reconnaissance, steps of soil exploration (boring, sampling, and testing), and the record of field exploration. Although different types of soil tests are discussed in this chapter, specific step-by-step procedures are outside the scope of this book. For specific step-by-step procedures, the reader is referred to one of the many soil testing manuals available, such as those of the American Society for Testing and Materials (ASTM) and the American Association of State Highway and Transportation Officials (AASHTO).

2-2 RECONNAISSANCE

A reconnaissance is a preliminary examination or survey of the job site. Usually, some useful information on the area (for example, maps or aerial photographs) will already be available, and an astute person can learn much about surface conditions and can get a general idea of subsurface conditions by simply visiting the site, observing thoroughly and carefully, and properly interpreting what is seen.

A first step in a preliminary soil survey of an area should be to collect and study any pertinent information that is already available. This could include general geological and topographical information available in the form of geological and topographic maps. Such maps are available from federal, state, and local governmental agencies (e.g., U.S. Geological Survey, Soil Conservation Service of the U.S. Department of Agriculture, and various state geological surveys).

Aerial photographs can provide geologic information over a large area. Proper interpretation of aerial photographs may reveal such information as land patterns, sinkhole cavities, landslides, surface drainage patterns, and the like. Such information can usually be obtained on a more widespread and thorough basis by aerial photography than by visiting the project site. Specific information on aerial photography is, however, beyond the scope of this book. For such information, the reader is referred to the many books available on aerial photo interpretation.

After carefully collecting and studying available pertinent information, the soils engineer or technologist should then visit the site in person, observe thoroughly and carefully, and interpret what is seen. The ability to do this successfully requires considerable practice and experience; however, a few generalizations are given next.

To begin with, significant information concerning surface conditions and general information about subsurface conditions in an area may be obtained by observing general topographic characteristics at the proposed job site and at nearby locations where subsurface soil is exposed. Such locations would include areas where the soil was cut or eroded (such as railroad and highway cuts, ditch and stream erosion, and quarries), thereby exposing subsurface soil strata.

The general topographic characteristics of an area can be of significance. For example, construction sites on swampy areas and dump areas, such as sanitary landfill, deserve particular attention in soil exploration.

Since the presence of water is an important consideration in working with soil and associated structures, several observations regarding water may be made during the reconnaissance. Groundwater levels may be noted by observing existing wells. Historical high water marks may be recorded on buildings, trees, and so on.

Often, valuable information can be obtained by talking with local inhabitants of the area. Such information could include flooding history, erosion patterns, mud slides, soil conditions, depth of overburden, groundwater level, and the like.

One final consideration with regard to reconnaissance is that the reconnoiterer should take numerous photographs of the area. Photographs of the proposed construction site, exposed subsurface strata, adjacent structures, and so on, can be invaluable in subsequent analysis and design processes and in later comparisons of conditions before and after construction.

The authors hope that the preceding discussion in this section has made the reader aware of the importance of reconnaissance with regard to soil exploration at a proposed construction site. In addition to providing important information, results of the reconnaissance help determine the necessary scope of subsequent soil exploration.

At some point prior to beginning the actual subsurface exploration (Sec. 2-3), it is important that underground utilities (water mains, sewer lines, etc.) be located. Location of such underground utilities is important in planning and carrying out subsequent subsurface exploration.

2-3 STEPS OF SOIL EXPLORATION

After obtaining all possible preliminary information as indicated in the preceding section, the next step is the actual subsurface soil exploration. This should be done by experienced personnel, using appropriate equipment. Much of soil mechanics practice can be successful only if one has long experience with which to compare each new problem.

Soil exploration may be thought of as consisting of three steps—boring, sampling, and testing. *Boring* refers to drilling or advancing a hole in the ground; *sampling* refers to removing soil from the hole; and *testing* refers to the determination of the characteristics or properties of the soil. These three steps appear simple in concept but are quite difficult in good practice and are discussed in detail in the remainder of this section.

Boring

Some of the more common types of borings are auger borings, wash borings, test pits, and core borings.

An *auger* (Fig. 2-1) is a screwlike tool used to bore a hole. Some augers are operated by hand and others are power operated. As the hole is bored a short distance, the auger may be lifted to remove soil. The removed soil may be used for field classification and laboratory testing, but it must not be considered as an undisturbed soil sample. It is difficult to use augers either in very soft clay or coarse sand because the hole tends to refill when the auger is removed. It may be difficult or impossible to use an auger below the water table because most saturated soils will not cling sufficiently to the auger for lifting. Hand augers may be used to bore to a depth of about 20 ft; power augers may be used to bore much deeper and quicker.

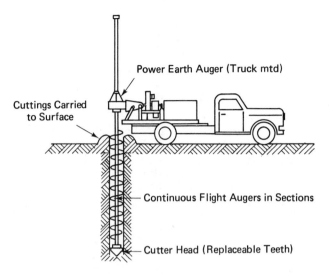

Power Earth Auger (Truck mtd)

Cuttings Carried
to Surface

Continuous Flight Augers in Sections

Cutter Head (Replaceable Teeth)

FIGURE 2-1   Auger boring. (Courtesy of Acker Drill Co.) [1][1]

Wash borings (Fig. 2-2) consist of simultaneous drilling and jetting action. To begin with, a casing is usually driven into the ground. A chopping bit attached to the end of a drilling rod (or wash pipe) is usually driven by hammer, thereby breaking up the soil in the casing. The jetting action is accomplished by pumping water downward through the drilling bit. The water emerges at the chopping bit and further serves to break up the soil. Returning water transports the soil to the ground surface, where samples can be collected for examination and classification purposes. Such samples are, of course, disturbed samples whose water content has been increased.

Test pits (Fig. 2-3) are excavated either manually or by equipment (backhoe or bulldozer). Test pits are generally bulky, expensive (when done manually), and limited to shallow exploration; nevertheless, they do have certain advantages. One is that, by observing the sides of the pit, the observer can inspect the soil in its natural condition. Another advantage is that they can be used to obtain undisturbed samples. (The student should note that even here the sample cannot be completely undisturbed.)

Core borings are used for drilling through rock. Such borings are often done using a diamond core drill in a core barrel sampler. The drill bit and core barrel sampler, attached to rods, are rotated by the drill. At the same time, water is circulated (pumped) through the rods and barrel, emerging at

---

[1] Reprinted with permission of Reston Publishing Company, Inc., A Prentice-Hall Company, 11480 Sunset Hills Road, Reston, Va. (This footnote applies to all succeeding citations to this reference in this book.)

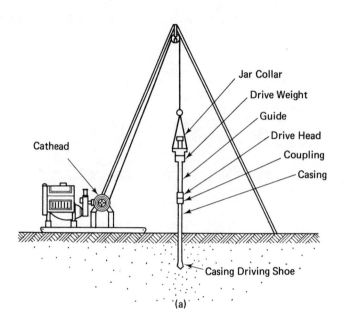

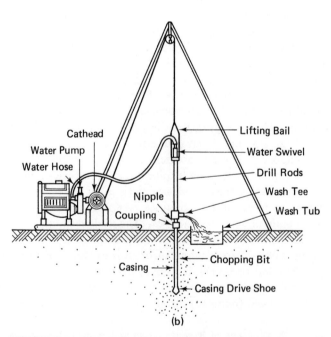

FIGURE 2-2 Typical setup for wash boring: (a) driving casing; (b) chopping and jetting. (Courtesy of Acker Drill Co.) [1]

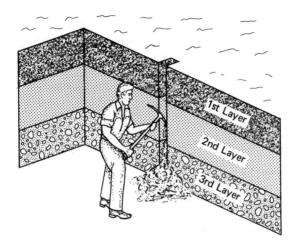

FIGURE 2-3 Test pit. [2]

the bit. The core remains in the core barrel and may be removed for examination by bringing the barrel to the surface.

The preceding paragraphs have discussed some of the more common types of borings. Once a means of boring has been decided upon, the question arises as to how many borings should be made. Obviously, the more borings that are made, the better the analysis of subsurface conditions should be. Borings are expensive, however, and a balance must be made between the cost of additional borings and the value of information gained from the additional borings.

As a rough guide for initial spacing of borings, the following are offered: for multistory buildings, 50 to 100 ft; for one-story buildings, earth dams, and borrow pits, 100 to 200 ft; and for highways (subgrade), 500 to 1000 ft. These spacings may be increased if soil conditions are found to be relatively uniform and they must be decreased if found to be nonuniform.

Once means of boring and spacing have been decided upon, the final question arises as to how deep the borings should be. In general, borings should extend through any unsuitable foundation strata (unconsolidated fill, organic soils, compressible layers such as soft fine grained soils, etc.) until soil of acceptable bearing capacity (hard or compact soil) is reached. If soil of acceptable bearing capacity is encountered at shallow depths, one or more borings should extend to such a depth to ensure that an underlying weaker layer, if found, will have a negligible effect on surface stability and settlement. In compressible fine-grained strata, borings should extend to a depth at which stress from superimposed load is so small that surface settlement is negligible.

In the case of very heavy structures, including tall buildings, borings in most cases should extend to bedrock. In all cases, it is advisable to investigate drilling at least one boring to bedrock.

The preceding discussion presented some general considerations regarding depths of borings. A more definitive criterion for determining the required minimum depth of test borings in cohesive soils is to carry the test boring down to a depth where the increase in stress due to the foundation loading (i.e., weight of the structure) is less than 10% of the effective overburden pressure. Figures 2-4, 2-5, and 2-6 were developed [3] to determine minimum depths of borings based on the 10% increase in stress criterion for cohesive soils. Figure 2-4 is for a continuous footing (such as a wall footing). Figure 2-5 is for a square footing with a design pressure between 1000 and 9000 psf, and Fig. 2-6 is for a square footing with a design pressure between 100 and 1000 psf. If the groundwater table is assumed to be at the base of the footing, the buoyant weight (submerged unit weight) of the soil should be used in these figures. If the groundwater table is lower than the distance *B* below the footing (*B* is the width of the footing), the wet unit weight of the soil should be used. For intermediate watertable conditions, an interpolation can be made between the required depths of boring for shallow and deep ground-water conditions, or the groundwater table can be conservatively assumed to be at the base of the footing. It should be noted that on the left side of Figs. 2-4 through 2-6 two scales are given, for footing width and minimum test

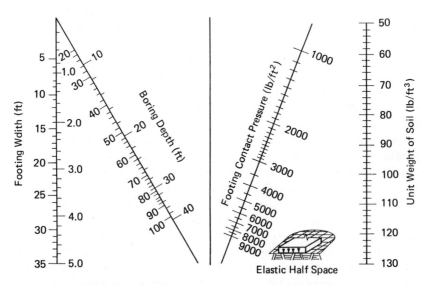

FIGURE 2-4 Infinite strip loading—Boussinesq-type solid. [3]

boring depth. In each figure the footing width given on one side of the scale corresponds with the boring depths given on the same side of the boring depth scale [3].

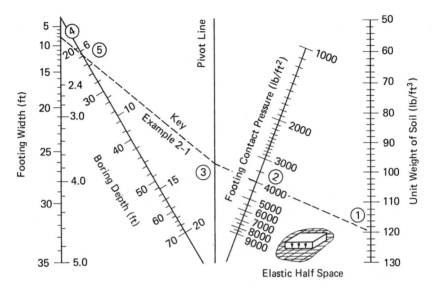

FIGURE 2-5   Square loading—Boussinesq-type solid. [3]

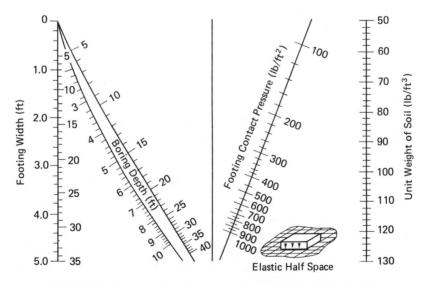

FIGURE 2-6   Square loading (low-pressure)—Boussinesq-type solid. [3]

EXAMPLE   2-1 [3]

Given

1. An 8-ft-square footing is subjected to a contact pressure of 4000 psf.
2. The wet unit weight of the soil supporting the footing is estimated to be 120 pcf.
3. The water table is estimated to be 30 ft beneath the footing.

Required

The minimum depth of test boring.

Solution

Since the water table is estimated to be 30 ft beneath the footing and the width of the footing is 8 ft, the wet unit weight of the soil should be used. From Fig. 2-5 with a wet unit weight of 120 pcf, contact pressure between footing and soil equal to 4000 psf, and a width of footing equal to 8 ft, the minimum depth of the boring is determined to be 22 ft.

Figures 2-4 through 2-6 are quite useful for estimating minimum required test boring depths in cohesive soils. In the final analysis, however, the depth of a specific boring should be determined by the soils engineer or technologist based on his or her knowledge, experience, judgment, and general knowledge of the specific area. Also, in some cases, depth (and spacing) of borings may be specified by local codes or company policy.

Sampling

Sampling refers to the method for taking soil or rock from bored holes. Samples may be classified as either "disturbed" or "undisturbed."

As mentioned previously in this section, in both auger borings and wash borings, soil is brought to the ground surface, where samples can be collected. Such samples are obviously *disturbed samples*, and thus some of their characteristics are changed. (Split-spoon samples, described in Sec. 2-4, also provide disturbed samples.) Disturbed samples should be placed in an airtight container (plastic bag or airtight jar, for example) and should, of course, be properly labeled as to date, location, bore-hole number, sampling depth, and so on. Disturbed samples are generally used for soil grain-size analysis, for determination of liquid limit and plastic limit, for determination of specific gravity of the soil, and for other tests, such as the compaction test or CBR (California bearing ratio) test.

For determination of certain other properties of soils, such as strength, compressibility, and permeability, it is necessary that the collected soil sample be exactly the same as it was when it existed in place within the ground. Such a soil sample is referred to as an *undisturbed sample*. It should be realized, however, that such a sample can never be completely undisturbed (i.e., be exactly the same as it was when it existed in place within the ground).

Undisturbed samples may be collected by several methods. If a test pit is available in a clay soil, an undisturbed sample may be obtained by simply carving a sample very carefully out of the side of the test pit. Such a sample should then be coated with paraffin wax and placed in an airtight container. This method is often considered to be too tedious, time consuming, and expensive to be done on a large scale, however.

The more common method of obtaining an undisturbed sample is to push a thin tube into the soil, thereby trapping the (undisturbed) sample inside the tube and then to remove the tube and sample intact. The ends of the tube should be sealed with paraffin wax immediately after the tube containing the sample is brought to the ground surface. The sealed tube should then be sent to the soils laboratory, where subsequent tests can be made on the sample, with the assumption that such test results are indicative of the properties of the soil as it existed in place within the ground.

The thin tube mentioned above is called a "Shelby tube." It is a 2- to 3-in.-diameter 16-gauge seamless steel tube, which is preferably pushed into the soil by static force rather than being driven by hammer.

Sampling of rock by core boring has been mentioned previously in this section.

Normally, samples (both disturbed and undisturbed) are collected at least every 5 ft in depth of the boring hole. If, however, any change in soil characteristics is noted within 5-ft intervals, additional samples should be made when a change is noted.

The importance of properly and accurately identifying and labeling each sample cannot be overemphasized.

After a boring has been made and samples taken, a very rough estimate of the groundwater level can be made. It is common practice to cover the hole (for example, with a small piece of plywood) for safety reasons, mark it for identification, leave it overnight, and return the next day to record the groundwater level. The hole should then be filled in to avoid subsequent injury to people or animals.

Testing

There exist a large number of tests that can be made on soil to determine various soil properties. These include both laboratory tests and field tests. Some of the most common tests are listed in Table 2-1. As indicated at the beginning of this chapter, the reader is referred to one of the many soil testing manuals available for specific step-by-step procedures involving these tests. Two of these tests—the standard penetration test and the vane test—are described in detail in Secs. 2-4 and 2-5.

TABLE 2-1 Common types of testing [2].

A. Laboratory Testing of Soils

| | | ASTM | | |
| | | | | |
Properties of Soil	*Type of Test*	*Designation of Standard Methods*	*Suggested Methods*	*AASHTO Designation*
Grain-size distribution	Mechanical analysis	D421, D422, D1140		T88
Consistency	Liquid limit (*LL*)	D423		T89
	Plastic limit (*PL*)	D424		T90
	Plasticity index (*PI*)	D424		T91
Unit weight	Specific gravity	D854		T100
Moisture	Natural water content			
	Field moisture equivalent	D426		T93
	Centrifuge moisture equivalent	D425		T94
Shear strength				
Cohesive soils	Unconfined compression		Yes	
Noncohesive soils	Direct shear		Yes	
General	Triaxial		Yes	
Volume change	Shrinkage factors	D427		T92
	Volume change		Yes	T116
	Expansion pressure		Yes	
Compressibility	Consolidation		Yes	
Permeability	Permeability		Yes	
Compaction characteristics	Standard proctor	D698		T99
	Modified proctor		Proposed 1958	T180
California bearing ratio (CBR)			Yes	

TABLE 2-1   (continued)

B.   Field Testing of Soils

| Properties of Soil | Type of Test | ASTM | | AASHTO Designation |
		Designation of Standard Methods	Suggested Methods	
Compaction control	Moisture-density relations	D698	Proposed	T99, T180
	In-place density	D1556	Yes	T147 T181
	Penetrometer needle		Proposed 1958	
Shear strength (soft clay)	Vane test			
Relative density (granular soil)	Penetration test			
Permeability	Pumping test			
Bearing capacity				
Pavements	CBR		Yes	
	Plate bearing	D1195 D1196		
Footings	Plate bearing	D1194		
Piles (vertical load)	Load test	D1143		
Batter piles	Latteral load test		Yes	

2-4   STANDARD PENETRATION TEST

The standard penetration test (SPT) is widely used in the United States. Relatively simple and inexpensive to perform, it is useful in determining certain properties of soils, particularly of cohesionless soils, for which undisturbed samples are not easily obtained.

The SPT utilizes a "split-spoon" sampler (Fig. 2-7). It is a 2-in.-O.D. $1\frac{3}{8}$-in.-I.D. tube, 18 to 24 in. long, that is split longitudinally down the middle. The split-spoon sampler is attached to the bottom of the drilling rod and driven into the soil with a drop hammer. Specifically, a 140-lb hammer falling 30 in. is used to drive the split-spoon sampler 18 in. into the soil. As the sampler is

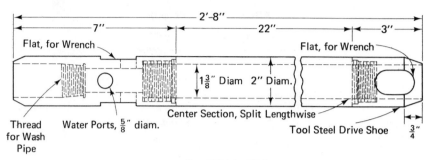

Total Weight 15 lb

FIGURE 2-7   Split-spoon sampler for the standard penetration test. [4]

driven the 18 in. into the soil, the number of blows required to penetrate each of the three 6-in. increments is recorded separately. The standard penetration resistance value (or N-value) is the number of blows required to penetrate the last 12 in. Thus, the N-value represents the number of blows per foot. After the blow counts have been obtained, the split-spoon sampler can be removed and opened (along the longitudinal split) to obtain a disturbed sample for subsequent examination and testing. [5]

SPT results (i.e., the N-values) are influenced by overburden pressure (effective weight of soil above) at the location where the blow count was made. Several methods have been proposed to correct N-values to reflect the influence of overburden pressure. Two of these methods are presented here.

One of these methods [6] utilizes the equation

$$C_N = 0.77 \log_{10} \frac{20}{\bar{p}} \tag{2-1}$$

where C_N = correction factor to be applied to the N-value recorded in the field

$\bar{p}$ = effective overburden pressure at the elevation of the SPT, tons/ft^2

This equation is not valid if $\bar{p}$ is less than 0.25 ton/ft^2. Figure 2-8 gives a graphical relationship, based in part on Eq. (2-1), for determining a correction factor to be applied to the N-value actually recorded in the field. If $\bar{p}$ is greater than or equal to 0.25 ton/ft^2, the correction factor may be determined using either Eq. (2-1) or Fig. 2-8. If $\bar{p}$ is less than 0.25 ton/ft^2, the correction factor should be taken from the figure.

$$\text{Correction factor } C_N = \frac{N_{\bar{p}=1}}{N_{\text{Field}}}$$

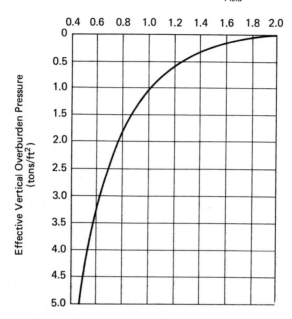

FIGURE 2-8 Chart for correction of *N*-values in sand for influence of overburden pressure (reference value of effective overburden pressure, 1 ton/ft²). [6]

The other method [7] utilizes the equations

$$N = \frac{4N'}{1 + 2p_0} \qquad \text{if } p_0 \leq 1.5 \text{ kips per square foot (ksf)} \qquad (2\text{-}2)$$

$$N = \frac{4N'}{3.25 + 0.5p_0} \qquad \text{if } p_0 \geq 1.5 \text{ ksf} \qquad (2\text{-}3)$$

where N = corrected *N*-value

         N' = *N*-value recorded in the field

         p_0 = effective overburden pressure, ksf

These two methods give comparable results. It will be noted that the first method [Eq. (2-1)] results in no adjustment of *N*-value at a depth where the effective overburden pressure is 1 ton/ft²; while the second method [Eqs. (2-2) and (2-3)] results in no adjustment of the *N*-value at a depth where the effective overburden pressure is 0.75 ton/ft² (1.5 ksf).

The reader is cautioned that, although the standard penetration test is widely used in the United States, the results are highly variable. Nevertheless, it is a useful guide in foundation analysis. Much experience is necessary to properly apply the results obtained. Outside the United States, other techniques are used. For example, in Europe the Dutch Cone is often preferred.

EXAMPLE 2-2

Given

> The SPT was performed at a depth of 20 ft in sand of unit weight 135 pcf. The blow count was 40.

Required

> The corrected N-value by each of the methods presented above.

Solution

1. By Eq. (2-1),

$$C_N = 0.77 \log_{10} \frac{20}{\bar{p}} \qquad (2\text{-}1)$$

$$\bar{p} = \frac{(20)(135)}{2000} = 1.35 \text{ tons/ft}^2$$

$$C_N = 0.77 \log_{10} \frac{20}{1.35}$$

$$= 0.90$$

(This value of 0.90 for C_N can also be obtained using Fig. 2-8 by entering 1.35 tons/ft² along the ordinate, moving horizontally to the curved line, and then moving upward to obtain the correction factor, C_N.)

Therefore, $N_{\text{corrected}} = (40)(0.90) = 36$.

2. By Eq. (2-2) or (2-3),

$$p_0 = \frac{(20)(135)}{1000} = 2.70 \text{ ksf}$$

Since $p_0 = 2.70 \text{ ksf} > 1.5 \text{ ksf}$, use Eq. (2-3).

$$N = \frac{4N'}{3.25 + 0.5p_0} \qquad (2\text{-}3)$$

$$= \frac{(4)(40)}{3.25 + (0.5)(2.70)}$$

$$N_{\text{corrected}} = 35$$

EXAMPLE 2-3

Given

Same data as given in Example 2-2 except that the water table is located at a depth of 5 ft below the ground surface.

Required

The corrected N-value by each of the methods presented above.

Solution

1. By Eq. (2-1),

$$C_N = 0.77 \log_{10} \frac{20}{\bar{p}} \tag{2-1}$$

$$\bar{p} = \frac{(5)(135) + (15)(135 - 62.4)}{2000} = 0.88 \text{ ton/ft}^2$$

$$C_N = 0.77 \log_{10} \frac{20}{0.88}$$

$$= 1.04$$

Therefore, $N_{\text{corrected}} = (40)(1.04) = 42$

2. By Eq. (2-2) or (2-3),

$$p_0 = \frac{(5)(135) + (15)(135 - 62.4)}{1000} = 1.76 \text{ ksf}$$

Since $p_0 = 1.76$ ksf > 1.5 ksf, use Eq. (2-3).

$$N = \frac{4N'}{3.25 + 0.5p_0} \tag{2-3}$$

$$= \frac{(4)(40)}{3.25 + (0.5)(1.76)}$$

$$N_{\text{corrected}} = 39$$

Through empirical testing, correlations between the (corrected) SPT N-value and several soil parameters have been established. These are particularly useful for cohesionless soils but are less reliable for cohesive soils. Table 2-2 gives correlations of relative density of cohesionless soil with SPT N-value and of state of consistency of cohesive soil with SPT N-value. Figure 2-9 gives a graphical relationship between angle of internal friction of cohesionless soil and SPT N-value. Figure 2-9 also gives graphical relationships between certain bearing capacity factors for cohesionless soil and SPT N-value. These will be utilized in Chap. 6.

TABLE 2-2   Penetration resistance and soil properties on basis of the standard penetration test [6].

SANDS (FAIRLY RELIABLE)		CLAYS (RATHER UNRELIABLE)	
Number of Blows per Foot N	*Relative Density*	*Number of Blows per Foot, N*	*Consistency*
		Below 2	Very soft
0–4	Very loose	2–4	Soft
4–10	Loose	4–8	Medium
10–30	Medium	8–15	Stiff
30–50	Dense	15–30	Very stiff
Over 50	Very dense	Over 30	Hard

2-5   VANE TEST [2]

The field vane test is a fairly simple test that can be used to determine in-place shear strength for soft clay soil—particularly those clay soils that lose part of their strength when disturbed (sensitive clays)—without taking an undisturbed sample. A vane tester (Fig. 2-10) is made up of two thin metal blades attached to a vertical shaft. The test is carried out by pushing the vane tester into the soil and then applying a torque to the vertical shaft. The cohesion of the clay can be computed using the formula [2, 8]

$$c = \frac{T}{\pi[(d^2h/2) + (d^3/6)]} \tag{2-4}$$

where c = cohesion of the clay, lb/ft^2

T = torque required to shear the soil, ft-lb

d = diameter of vane tester, ft

h = height of vane tester, ft

Since shear strength of a cohesive soil results primarily from cohesion, Eq. (2-4) gives, in effect, the shear strength of the soil.

It should be emphasized that the field vane test is suitable only for use in soft or sensitive clay. Also, no soil sample is obtained for subsequent examination and testing when a field vane test is made.

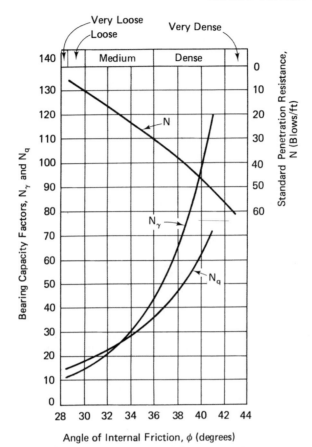

FIGURE 2-9 Curves showing the relationship between bearing capacity factors and ϕ, as determined by theory, and rough empirical relationship between bearing capacity factors or ϕ and values of standard penetration resistance N. [6]

2-6 RECORD OF SOIL EXPLORATION

It is of the utmost importance that complete and accurate records be kept of all data collected. Boring, sampling, and testing are often costly undertakings, and to fail to keep good, accurate records is not only senseless, but it may also be dangerous.

To begin with, a good map giving the specific location of each boring should be available. Each boring should be identified (by number, for example), and the location of each boring should be documented by measurement to permanent features. Such a map is illustrated in Fig. 2-11.

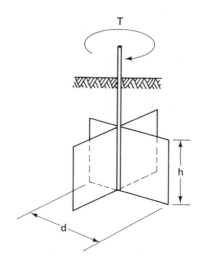

FIGURE 2-10 Vane tester. [2]

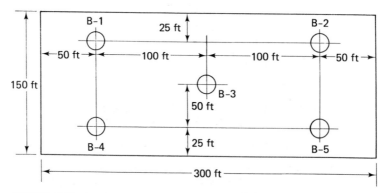

FIGURE 2-11 Example map showing boring locations on 150-ft by 300-ft construction site.

For each boring that is made, all pertinent data should be recorded in the field on a boring log sheet. Normally, these would be preprinted forms containing blanks for filling in appropriate data. An example of a boring log is given in Fig. 2-12.

The soil data obtained from a series of test borings can best be presented by preparing a geologic profile. Such a profile shows the arrangement of the various layers of soil as well as the groundwater level, existing and proposed structures, and soil properties data (SPT, for example). Each bore hole is identified and indicated on the geologic profile by a vertical line. An example of a geologic profile is shown in Fig. 2-13.

A geologic profile is prepared by indicating on each bore hole on the

ABC DRILLING COMPANY, INC.
NEWARK, NEW YORK

BORING NO. 5
ORD. ELEV. 372.4

PROJECT: Job No. 459

Name Eureka Warehouse

Address Illion, New York

CASING (Size & Type) 2½" Drive Pipe

SAMPLE SPOON (Size & Type) 2" O.D.S.S.

HAMMER (Csg): Wt. 250 lbs., Drop 24 in.

(Spoon): Wt. 140 lbs., Drop 30 in.

DATE: Started 7/28/-- Completed 7/29/-- Driller Henry James

GROUND WATER OBSERVATIONS

Date	Time	Depth	Casing at
7/29/--	3:00 PM	18'3''	15'0''
''	4:00 PM	12'0''	10'0''
''	4:30 PM	8'0''	5'0''
7/30/--	8:30 AM	7'0''	Out

DEPTH FT.	BLOWS CSG.	BLOWS SPOON	N	Samples	
0					Black and grey moist FILL: cinders, brick and silt
1	2				
2	16	11 / 8	12	S #1, 1'-2'6''	
3	9	4			3'0''
4	3	1 / 1	3	S #2, 3'-4'6''	
5	3	2			Black PEAT
6	3	P / 1	2	S #3, 5'-6'6''	6'0''
7	5	1			
8	6	3 / 6	11	S #4, 7'-8'6''	
9	8	5			Grey moist SILT with embedded fine gravel, trace of fine sand
10	9				
11	3	4 / 8	14	S #5, 10'-11'6''	
12	8	6			12'6''
13	15	15			Weathered SHALE
14	32	18 / 21	39	S #6, 12'6''-14'	TOP OF ROCK 15'0''
15	78				
16				Core Boring Series M- double tube core barrel, 2 in. diam. bit.	Weathered grey SHALE Run #1, 15'0'' - 20'0'' Recovered 30'' - 50% Lost water @ 16'6''
17					
18					
19					
20					20'0''
21					SHALE and SANDSTONE Run #2, 20'0'' - 25'0'' Recovered 56'' - 93% Steady resistance while drilling
22					
23					
24					
25					25'0''

FIGURE 2-12 Boring log. [9]

51

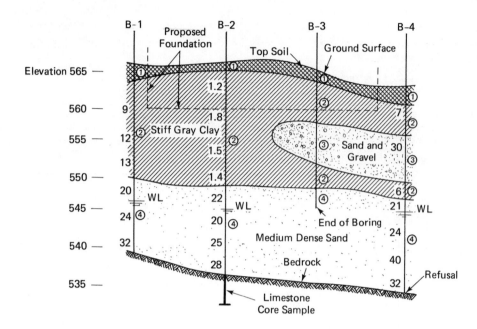

FIGURE 2-13 Example of geologic profile. [2]

profile (i.e., each vertical line representing a bore hole) the data obtained by boring, sampling, and testing. From these data, the soil layers can be sketched in. Obviously, the more bore holes and the closer they are spaced, the more accurate the resulting geologic profile will be.

2-7 CONCLUSION

The student should consider this chapter to be one of the most important in this book. Analysis of soil and design of associated structures are of questionable value if the soil exploration data are not accurately determined and reported.

The authors hope that this chapter will give the student an effective introduction to actual soil exploration. However, learning to conduct soil exploration well requires much practice and varied experience under the guidance of experienced practitioners. Not only is it a complex science, it is a difficult art.

References

[1] DAVID F. MCCARTHY, *Essentials of Soil Mechanics and Foundations*, Reston Publishing Company, Inc., Reston, Va., 1977.

[2] WAYNE C. TENG, *Foundation Design*, Prentice-Hall, Inc., Englewood Cliffs, N.J., 1962.

[3] RICHARD D. BARKSDALE AND MILTON O. SCHREIBER, "Calculating Test-boring Depths," *Civil Engineering, ASCE*, **49** (8), 74–75 (1979).

[4] R. H. KAROL, *Soils and Soil Engineering*, Prentice-Hall, Inc., Englewood Cliffs, N.J., 1960.

[5] KARL TERZAGHI AND RALPH B. PECK, *Soil Mechanics in Engineering Practice*, John Wiley & Sons, Inc., New York, 1967. Copyright © 1967, John Wiley & Sons, Inc. Reprinted by permission of John Wiley & Sons, Inc.

[6] RALPH B. PECK, WALTER E. HANSEN, AND THOMAS H. THORNBURN, *Foundation Engineering*, 2nd ed., John Wiley & Sons, Inc., New York, 1974. Copyright © 1974, by John Wiley & Sons, Inc. Reprinted by permission of John Wiley & Sons, Inc.

[7] ABDEL RAHMAN SADIK SAID BAZARAA, "Use of the Standard Penetration Test for Estimating Settlements of Shallow Foundations on Sand," Ph.D. thesis, University of Illinois, 1967.

[8] A. W. SKEMPTON AND A. W. BISHOP, "The Measurement of the Shear Strength of Soils," *Geotechnique* **2** (2) (1950).

[9] B. K. HOUGH, *Basic Soils Engineering*, 2nd ed., The Ronald Press Company, New York, 1969. Copyright © 1969, by John Wiley & Sons, Inc. Reprinted by permission of John Wiley & Sons, Inc.

3

Stress Distribution in Soil

3-1 INTRODUCTION

If a vertical load of 1 ton is applied to a column of 1 ft² cross-sectional area, and the column rests directly on a soil surface, the vertical pressure exerted by the column onto the soil would be, on the average, 1 ton/ft² (neglecting the weight of the column). In addition to this pressure at the area of contact between column and soil, the stress influence extends both downward and outward within the soil in the general area where the load is applied. The increase in pressure in the soil at any horizontal plane below the load is greatest directly under the load and diminishes outwardly (see Fig. 3-1). The magnitude of the pressure decreases with increasing depth. This is illustrated in Fig. 3-1, where pressure p_2 at depth d_2 is less than pressure p_1 at depth d_1. Figure 3-1 also illustrates the increase in the area of stress influence outward with increase in depth.

This stress distribution in soil is quite important to the soils engineer and technologist—particularly with regard to stability analysis and settlement analysis of foundations. The remainder of this chapter deals with the quantitative analysis of stress distribution in soil.

3-2 VERTICAL PRESSURE BELOW A CONCENTRATED LOAD

Two methods for calculating pressure below a concentrated load are presented here—the Westergaard equation and the Boussinesq equation. Both of these equations result from the theory of elasticity, which assumes that

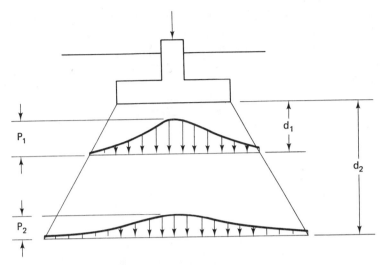

FIGURE 3-1 Distribution of pressure.

stress is proportional to strain. Implicit in this assumption is a homogeneous material, and soil is very seldom a homogeneous material. The Westergaard equation is based on alternating thin layers of an elastic material between layers of an inelastic material. The Boussinesq equation assumes a homogeneous soil throughout. [1]

Westergaard Equation [1, 2]

The Westergaard equation is as follows:

$$q = \frac{Q\sqrt{(1 - 2\mu)/(2 - 2\mu)}}{2\pi z^2[(1 - 2\mu)/(2 - 2\mu) + (r/z)^2]^{3/2}} \tag{3-1}$$

where q = vertical stress at depth z

Q = concentrated load

μ = Poisson's ratio (ratio of the strain in a material in a direction normal to an applied stress to the strain parallel to the applied stress)

z = depth

r = horizontal distance from point of application of Q to point at which q is desired

Note: q, the vertical stress at depth z resulting from load Q, is sometimes referred to as the *vertical stress increment*, since it represents stress added by the load to the stress existing prior to application of the load. (The stress existing prior to application of the load is the overburden pressure.) This equation gives stress q as a function of both the vertical distance z and the horizontal distance r between the point of application of Q and the point at which q is desired (see Fig. 3-2). If Poisson's ratio is taken to be zero, Eq. (3-1) reduces to

$$q = \frac{Q}{\pi z^2 [1 + 2(r/z)^2]^{3/2}} \qquad (3\text{-}2)$$

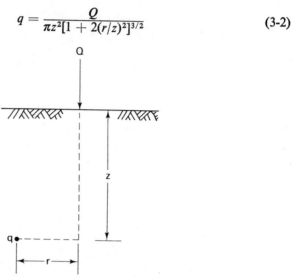

FIGURE 3-2

Boussinesq Equation [1]

The Boussinesq equation is as follows:

$$q = \frac{3Q}{2\pi z^2 [1 + (r/z)^2]^{5/2}} \qquad (3\text{-}3)$$

where the terms are the same as those in Eq. (3-1). This equation also gives stress q as a function of both the vertical distance z and the horizontal distance r. For low r/z ratios, the Boussinesq equation gives higher values of q than the Westergaard equation and is more widely used.

Examples 3-1 and 3-2 illustrate the use of the Boussinesq equation to calculate vertical stress below a concentrated load.

EXAMPLE 3-1

Given

A concentrated load of 250 tons is applied to the ground surface.

Required

The vertical stress increment in psf due to this load at a depth of 20 ft directly below the load.

Solution

From Eq. (3-3),

$$q = \frac{3Q}{2\pi z^2[1 + (r/z)^2]^{5/2}} \tag{3-3}$$

From given, $z = 20$ ft

$r = 0$

$Q = 250$ tons $= 500{,}000$ lb

$$q = \frac{(3)(500{,}000)}{(2)(\pi)(20)^2[1 + (0/20)^2]^{5/2}}$$

$$= 597 \text{ psf}$$

EXAMPLE 3-2

Given

A concentrated load of 250 tons is applied to the ground surface.

Required

The vertical stress increment in psf due to this load at a point located at a depth of 20 ft below the ground surface and at a horizontal distance of 16 ft from the line of the concentrated load (i.e., $r = 16$ ft, $z = 20$ ft, as illustrated in Fig. 3-3).

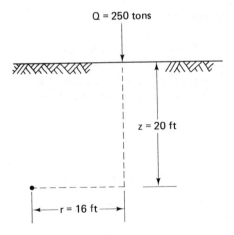

FIGURE 3-3

Solution

From Eq. (3-3),

$$q = \frac{3Q}{2\pi z^2[1 + (r/z)^2]^{5/2}} \tag{3-3}$$

From given, $z = 20$ ft

$$r = 16 \text{ ft}$$

$$Q = 250 \text{ tons} = 500{,}000 \text{ lb}$$

$$q = \frac{(3)(500{,}000)}{(2)(\pi)(20)^2[1 + (16/20)^2]^{5/2}}$$

$$= 173 \text{ psf}$$

3-3 VERTICAL PRESSURE BELOW A LOADED SURFACE AREA (UNIFORM LOAD)

Methods presented in Sec. 3-2 deal with the determination of vertical pressure below a concentrated load. Usually, however, concentrated loads are not applied directly onto the soil. Instead, concentrated loads rest on footings, piers, and the like, and the load is applied to the soil through the footings or piers in the form of a loaded surface area (uniform load). Analysis of stress distributions resulting from loaded surface areas is generally more complicated than those resulting from concentrated loads.

Two methods for computing vertical pressure below a loaded surface area will be discussed here. One of them is called the "approximate method," and the other is based on elastic theory.

Approximate Method [1]

The approximate method is based on the assumption that the area (in a horizontal plane) of stress below a concentrated load increases with depth as shown in Fig. 3-4. With the 2:1 slope shown, it is apparent that at any depth z, both L and B are increased by the amount z. Accordingly, stress at depth z is given by

$$q = \frac{Q}{(B + z)(L + z)} \tag{3-4}$$

where q = approximate vertical stress at depth z

Q = total load

B = width

L = length

z = depth

Since Q, L, and B are constants for a given application, it is obvious that the stress at depth z (q) decreases as depth increases. This method should be con-

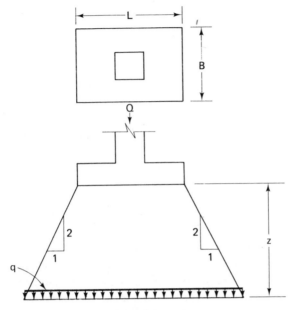

FIGURE 3-4 [1]

sidered, at best, as being crude. It may be useful for preliminary stability analysis of footings; however, for settlement analysis the approximate method may likely not be accurate enough, and a more accurate approach based on elastic theory (discussed later in this section) may be required.

Example 3-3 illustrates the use of the approximate method to calculate vertical pressure below a uniform load.

EXAMPLE 3-3

Given

A 10-ft by 15-ft rectangular area carrying a uniform load of 5000 lb/ft² is applied to the ground surface.

Required

The vertical stress (the vertical unit pressure) increment due to this load at a depth of 20 ft below the ground surface by the approximate method.

Solution

From Eq. (3-4),

$$q = \frac{Q}{(B + z)(L + z)} \tag{3-4}$$

$$Q = (5000)(10)(15) = 750,000 \text{ lb}$$

$$B = 10 \text{ ft}$$

$$L = 15 \text{ ft}$$

$$z = 20 \text{ ft}$$

$$q = \frac{750{,}000}{(10 + 20)(15 + 20)} = 714 \text{ psf}$$

Method Based on Elastic Theory

Uniform load on a circular area   Vertical pressure below a uniform load on a circular area can be determined utilizing Table 3-1. In the table, z and r represent, respectively, the depth and radial horizontal distance from the center of the circle to the point at which pressure is desired (these are similar to the z and r shown in Fig. 3-2); and "a" represents the radius of the circle on which the uniform loads acts. To calculate vertical pressure below a uniform load on a circular area, the ratios z/a and r/a are computed and an "influence coefficient" is determined from Table 3-1. This influence coefficient is simply multiplied by the uniform load applied to the circular area to determine the pressure at the desired point. Example 3-4 illustrates this method.

TABLE 3-1   Influence coefficients for points under uniform loaded circular area [3].

z/a (1)	0 (2)	0.25 (3)	0.50 (4)	1.0 (5)	1.5 (6)	2.0 (7)	2.5 (8)	3.0 (9)	3.5 (10)	4.0 (11)
0.25	0.986	0.983	0.964	0.460	0.015	0.002	0.000	0.000	0.000	0.000
0.50	0.911	0.895	0.840	0.418	0.060	0.010	0.003	0.000	0.000	0.000
0.75	0.784	0.762	0.691	0.374	0.105	0.025	0.010	0.002	0.000	0.000
1.00	0.646	0.625	0.560	0.335	0.125	0.043	0.016	0.007	0.003	0.000
1.25	0.524	0.508	0.455	0.295	0.135	0.057	0.023	0.010	0.005	0.001
1.50	0.424	0.413	0.374	0.256	0.137	0.064	0.029	0.013	0.007	0.002
1.75	0.346	0.336	0.309	0.223	0.135	0.071	0.037	0.018	0.009	0.004
2.00	0.284	0.277	0.258	0.194	0.127	0.073	0.041	0.022	0.012	0.006
2.5	0.200	0.196	0.186	0.150	0.109	0.073	0.044	0.028	0.017	0.011
3.0	0.146	0.143	0.137	0.117	0.091	0.066	0.045	0.031	0.022	0.015
4.0	0.087	0.086	0.083	0.076	0.061	0.052	0.041	0.031	0.024	0.018
5.0	0.057	0.057	0.056	0.052	0.045	0.039	0.033	0.027	0.022	0.018
7.0	0.030	0.030	0.029	0.028	0.026	0.024	0.021	0.019	0.016	0.015
10.00	0.015	0.015	0.014	0.014	0.013	0.013	0.013	0.012	0.012	0.011

EXAMPLE 3-4

Given

1. A circular area carrying a uniformly distributed load of 2000 psf is applied to the ground surface.
2. The radius of the circular area is 10 ft.

Required

The vertical stress increment in psf due to this uniform load:

1. At a point 20 ft below the center of the circular area.
2. At a point 20 ft below the ground surface at a horizontal distance of 5 ft from the center of the circular area (i.e., $r = 5$ ft, $z = 20$ ft).
3. At a point 20 ft below the edge of the circular area.
4. At a point 20 ft below the ground surface at a horizontal distance of 18 ft from the center of the circular area (i.e., $r = 18$ ft, $z = 20$ ft).

Solution

1. q = influence coefficient multiplied by the uniform load

   with $a = 10$ ft (radius of circle)

   $r = 0$ ft

   $z = 20$ ft

$$\frac{z}{a} = \frac{20}{10} = 2.00$$

$$\frac{r}{a} = \frac{0}{10} = 0$$

The influence coefficient from Table 3-1 $= 0.284$

$$q = (0.284)(2000) = 568 \text{ psf}$$

2. with $a = 10$ ft

   $r = 5$ ft

   $z = 20$ ft

$$\frac{z}{a} = \frac{20}{10} = 2.00$$

$$\frac{r}{a} = \frac{5}{10} = 0.5$$

The influence coefficient from Table 3-1 = 0.258

$$q = (0.258)(2000) = 516 \text{ psf}$$

3. with $a = 10$ ft

$r = 10$ ft

$z = 20$ ft

$$\frac{z}{a} = \frac{20}{10} = 2.00$$

$$\frac{r}{a} = \frac{10}{10} = 1.00$$

The influence coefficient from Table 3-1 = 0.194

$$q = (0.194)(2000) = 388 \text{ psf}$$

4. with $a = 10$ ft

$r = 18$ ft

$z = 20$ ft

$$\frac{z}{a} = \frac{20}{10} = 2.00$$

$$\frac{r}{a} = \frac{18}{10} = 1.8$$

From Table 3-1:

when $\dfrac{z}{a} = 2.00$ and $\dfrac{r}{a} = 1.5$,   influence coefficient $= 0.127$

when $\dfrac{z}{a} = 2.00$ and $\dfrac{r}{a} = 2.00$,   influence coefficient $= 0.073$

By interpolation between 0.127 and 0.073, the desired coefficient for $z/a = 2.00$ and $r/a = 1.8$ is

$$0.073 + \left(\frac{0.127 - 0.073}{5}\right)(2) = 0.095$$

or

$$0.127 - \left(\frac{0.127 - 0.073}{5}\right)(3) = 0.095$$

$$q = (0.095)(2000) = 190 \text{ psf}$$

Uniform load on a rectangular area   Vertical pressure below a uniform load on a rectangular area can be determined utilizing Table 3-2. In the table, z,

TABLE 3-2 Influence coefficients for rectangular areas [3, 4].

$m = A/z$ or $n = B/z$	\multicolumn{18}{c}{$n = B/z$ or $m = A/z$}																	
	0.1	0.2	0.3	0.4	0.5	0.6	0.7	0.8	0.9	1.0	1.2	1.5	2.0	2.5	3.0	5.0	10.0	∞
0.1	0.005	0.009	0.013	0.017	0.020	0.022	0.024	0.026	0.027	0.028	0.029	0.030	0.031	0.031	0.032	0.032	0.032	0.032
0.2	0.009	0.018	0.026	0.033	0.039	0.043	0.047	0.050	0.053	0.055	0.057	0.059	0.061	0.062	0.062	0.062	0.062	0.062
0.3	0.013	0.026	0.037	0.047	0.056	0.063	0.069	0.073	0.077	0.079	0.083	0.086	0.089	0.090	0.090	0.090	0.090	0.090
0.4	0.017	0.033	0.047	0.060	0.071	0.080	0.087	0.093	0.098	0.101	0.106	0.110	0.113	0.115	0.115	0.115	0.115	0.115
0.5	0.020	0.039	0.056	0.071	0.084	0.095	0.103	0.110	0.116	0.120	0.126	0.131	0.135	0.137	0.137	0.137	0.137	0.137
0.6	0.022	0.043	0.063	0.080	0.095	0.107	0.117	0.125	0.131	0.136	0.143	0.149	0.153	0.155	0.156	0.156	0.156	0.156
0.7	0.024	0.047	0.069	0.087	0.103	0.117	0.128	0.137	0.144	0.149	0.157	0.164	0.169	0.170	0.171	0.172	0.172	0.172
0.8	0.026	0.050	0.073	0.093	0.110	0.125	0.137	0.146	0.154	0.160	0.168	0.176	0.181	0.183	0.184	0.185	0.185	0.185
0.9	0.027	0.053	0.077	0.098	0.116	0.131	0.144	0.154	0.162	0.168	0.178	0.186	0.192	0.194	0.195	0.196	0.196	0.196
1.0	0.028	0.055	0.079	0.101	0.120	0.136	0.149	0.160	0.168	0.175	0.185	0.193	0.200	0.202	0.203	0.204	0.205	0.205
1.2	0.029	0.057	0.083	0.106	0.126	0.143	0.157	0.168	0.178	0.185	0.196	0.205	0.212	0.215	0.216	0.217	0.218	0.218
1.5	0.030	0.059	0.086	0.110	0.131	0.149	0.164	0.176	0.186	0.193	0.205	0.215	0.223	0.226	0.228	0.229	0.230	0.230
2.0	0.031	0.061	0.089	0.113	0.135	0.153	0.169	0.181	0.192	0.200	0.212	0.223	0.232	0.236	0.238	0.239	0.240	0.240
2.5	0.031	0.062	0.090	0.115	0.137	0.155	0.170	0.183	0.194	0.202	0.215	0.226	0.236	0.240	0.242	0.242	0.244	0.244
3.0	0.032	0.062	0.090	0.115	0.137	0.156	0.171	0.184	0.195	0.203	0.216	0.228	0.238	0.242	0.244	0.244	0.246	0.247
5.0	0.032	0.062	0.090	0.115	0.137	0.156	0.172	0.185	0.196	0.204	0.217	0.229	0.239	0.244	0.246	0.249	0.249	0.249
10.0	0.032	0.062	0.090	0.115	0.137	0.156	0.172	0.185	0.196	0.205	0.218	0.230	0.240	0.244	0.247	0.249	0.250	0.250
∞	0.032	0.062	0.090	0.115	0.137	0.156	0.172	0.185	0.196	0.205	0.218	0.230	0.240	0.244	0.247	0.249	0.250	0.250

A, and *B* represent, respectively, the depth below the loaded surface and the width and length of the rectangle on which the uniform load acts. To calculate vertical pressure below a uniform load on a rectangular area, the ratios $n = B/z$ and $m = A/z$ are computed and an "influence coefficient" is determined from Table 3-2. Either m or n can be read along the first column and the other one (n or m) is read across the top. The influence coefficient can also be determined utilizing Fig. 3-5. The influence coefficient is simply multiplied by the uniform load applied to the rectangular area to determine the pressure at depth z below each corner of the rectangle. Example 3-5 illustrates this method.

EXAMPLE 3-5

Given

A 15-ft by 20-ft rectangular foundation carrying a uniform load of 4000 psf is applied to the ground surface.

Required

The vertical stress increment in psf due to this uniform load at a point 10 ft below the corner of the rectangular loaded area. Use the influence coefficient method.

Solution

From Fig. 3-5, with

$$A = mz = 15 \text{ ft}; \quad z = 10 \text{ ft} \qquad m = \frac{15}{10} = 1.5$$

$$B = nz = 20 \text{ ft}; \quad z = 10 \text{ ft} \qquad n = \frac{20}{10} = 2.0$$

The influence coefficient $= 0.2235$

$$q = \text{influence coefficient multiplied by the uniform load}$$
$$= (0.2235)(4000) = 894 \text{ psf}$$

It should be emphasized that the pressure determined from the influence coefficients utilizing Table 3-2 or Fig. 3-5 (as in Example 3-5) is acting at depth z *directly below a corner of the rectangular area*. This is shown in Fig. 3-6, where such a computed stress acts at point *C*. It is sometimes necessary to determine the pressure below a rectangular loaded area at points other than directly below a corner of the rectangular area. For example, it may be necessary to determine the pressure at some depth directly below the center of a rectangular area, or at some point outside the downward projection of the rectangular area. This can be accomplished by dividing the area into rectangles, each of which has one corner directly above the point at which the

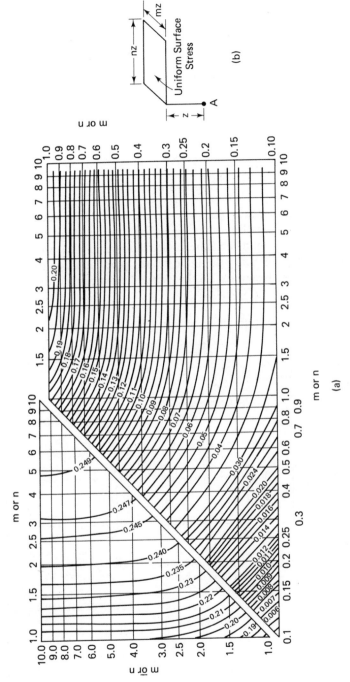

FIGURE 3-5 Chart for use in determining vertical stresses below corners of loaded rectangular surface areas on elastic, isotropic materials. [5, 6]

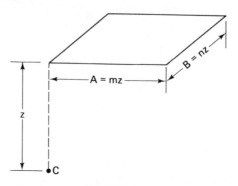

FIGURE 3-6

pressure is desired at depth z. The pressure is computed for each rectangle in the usual manner and the results added or subtracted to get the total pressure. Figure 3-7 should facilitate understanding of this procedure. In each case of Fig. 3-7, the heavy dot indicates the point at which the pressure at depth z is required. Examples 3-6 through 3-8 illustrate this procedure.

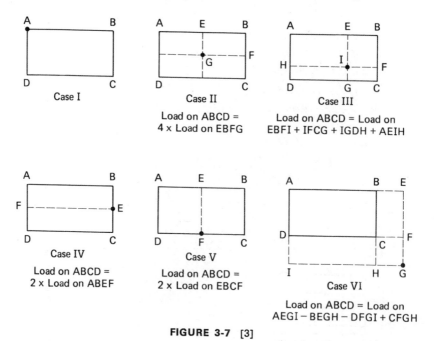

FIGURE 3-7 [3]

EXAMPLE 3-6

Given

A 20-ft by 30-ft rectangular foundation carrying a uniform load of 6000 psf is applied to the ground surface.

Required

The vertical stress increment in psf due to this uniform load at a depth of 20 ft below the center of the loaded area. (See point A in Fig. 3-8.)

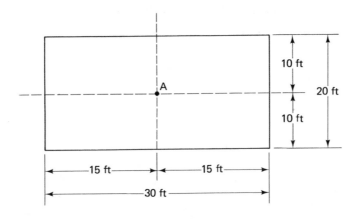

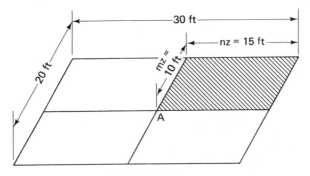

FIGURE 3-8

Solution

This corresponds to case II of Fig. 3-7, so the area is divided into four equal parts.

$$A = mz = 10 \text{ ft}; \quad z = 20 \text{ ft} \quad m = \frac{10}{20} = 0.5$$

$$B = nz = 15 \text{ ft}; \quad z = 20 \text{ ft} \quad n = \frac{15}{20} = 0.75$$

From Fig. 3-5, the influence coefficient = 0.1070 for a 10-ft by 15-ft loaded area. Since the original area of 20 ft by 30 ft consists of four smaller equal areas of 10 ft by 15 ft and each of these four areas shares a corner at point A,

$$q = (4)(0.1070)(6000) = 2570 \text{ psf}$$

EXAMPLE 3-7

Given

1. An L-shaped building (in plan) shown in Fig. 3-9.
2. The load exerted by the structure is 1400 psf.

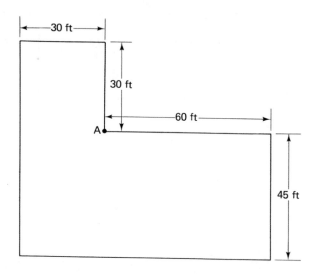

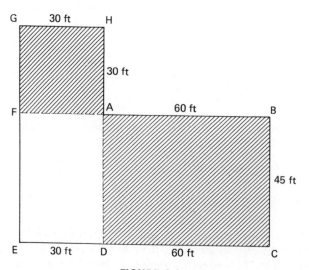

FIGURE 3-9

Required

Determine the vertical stress increment in psf due to the structure load at a depth of 15 ft below the interior corner A of the L-shaped building. Assume that the foundation is under the entire building.

Solution

Divide the L-shaped building into three smaller areas, $ABCD$, $ADEF$, and $AFGH$. Note that these three areas share a common corner at point A (or corner A).

Area ABCD

From Fig. 3-5,

$$A = mz = 60 \text{ ft}; \quad z = 15 \text{ ft} \quad m = \frac{60}{15} = 4$$

$$B = nz = 45 \text{ ft}; \quad z = 15 \text{ ft} \quad n = \frac{45}{15} = 3$$

The influence coefficient $= 0.2454$.

Area ADEF

From Fig. 3-5,

$$A = mz = 30 \text{ ft}; \quad z = 15 \text{ ft} \quad m = \frac{30}{15} = 2$$

$$B = nz = 45 \text{ ft}; \quad z = 15 \text{ ft} \quad n = \frac{45}{15} = 3$$

The influence coefficient $= 0.2375$.

Area AFGH

From Fig. 3-5,

$$A = mz = 30 \text{ ft}; \quad z = 15 \text{ ft} \quad m = \frac{30}{15} = 2$$

$$B = nz = 30 \text{ ft}; \quad z = 15 \text{ ft} \quad n = \frac{30}{15} = 2$$

The influence coefficient $= 0.2325$.

$$q = \sum \text{influence coefficients multiplied by the uniform load}$$

$$= (0.2454 + 0.2375 + 0.2325)(1400)$$

$$= 1000 \text{ psf}$$

EXAMPLE 3-8

Given

1.   A rectangular loaded area *ABCD* shown in plan in Fig. 3-10.
2.   The load exerted on the area is 1600 psf.

Required

Find the vertical stress increment in psf due to the exerted load at a depth of 10 ft below point *G* (see Fig. 3-10). Use influence coefficients.

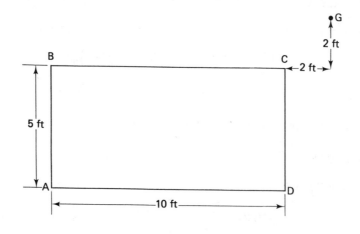

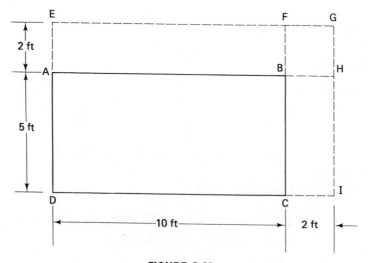

FIGURE 3-10

Solution

This corresponds to case VI of Fig. 3-7. The influence coefficient for the vertical stress increment under point G due to the uniform load on the area $ABCD$ may be obtained from the coefficients for various rectangles as follows:

$$\text{Load on } ABCD = \text{load on } DEGI - AEGH - CFGI + BFGH$$

(*Note:* In the equation above, the last term $BFGH$ is added because when $AEGH$ is subtracted, the area $BFGH$ is included in it and when $CFGI$ is subtracted, the area $BFGH$ is also included in it. Thus, the effect of area $BFGH$ has been subtracted twice. Thus, it must be added in order that its effect be subtracted only one time.)

Area DEGI

From Fig. 3-5,

$$A = mz = 7 \text{ ft}; \quad z = 10 \text{ ft} \qquad m = \frac{7}{10} = 0.7$$

$$B = nz = 12 \text{ ft}; \quad z = 10 \text{ ft} \qquad n = \frac{12}{10} = 1.2$$

The influence coefficient for area $DEGI = 0.157$.

Area AEGH

From Fig. 3-5,

$$A = mz = 2 \text{ ft}; \quad z = 10 \text{ ft} \qquad m = \frac{2}{10} = 0.2$$

$$B = nz = 12 \text{ ft}; \quad z = 10 \text{ ft} \qquad n = \frac{12}{10} = 1.2$$

The influence coefficient for area $AEGH = 0.057$.

Area CFGI

From Fig. 3-5,

$$A = mz = 7 \text{ ft}; \quad z = 10 \text{ ft} \qquad m = \frac{7}{10} = 0.7$$

$$B = nz = 2 \text{ ft}; \quad z = 10 \text{ ft} \qquad n = \frac{2}{10} = 0.2$$

The influence coefficient for area $CFGI = 0.047$.

Area BFGH

From Fig. 3-5,

$$A = mz = 2 \text{ ft}; \quad z = 10 \text{ ft} \qquad m = 0.2$$

$$B = nz = 2 \text{ ft}; \quad z = 10 \text{ ft} \qquad n = 0.2$$

The influence coefficient for area $BFGH = 0.018$.

$$q = (0.157 - 0.057 - 0.047 + 0.018)(1600)$$
$$= 114 \text{ psf}$$

Another method for computing vertical pressure below a uniform load on any area is by using an influence chart (Fig. 3-11) developed by Newmark [6] based on Boussinesq's equation. To utilize this method, a sketch (plan view) must be made of the loaded area on tracing paper drawn to such a scale

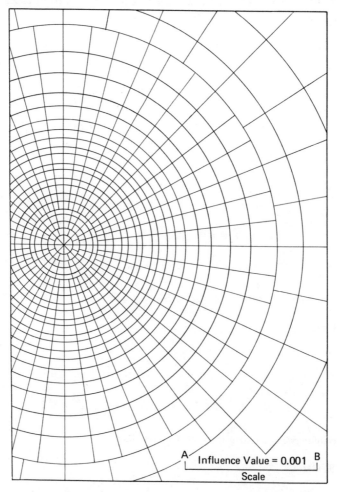

FIGURE 3-11 Newmark influence chart for computing vertical pressure. (After Corps of Engineers.) [7]

that the distance *AB* on Fig. 3-11 equals the depth at which the pressure is desired. This sketch is placed on the chart (Fig. 3-11) so that the point below which the pressure is desired coincides with the center of the chart. The next step is to count the quasi rectangles enclosed by the loaded area. The pressure at the indicated point at the desired depth is determined by multiplying the number of quasi rectangles by the applied uniform load by 0.001. As indicated on Fig. 3-11, the number 0.001 is the "influence value" for this particular chart. The same sketch may be used to determine the pressure at other points at the same depth by shifting the sketch until the desired point coincides with the center of the chart and counting the quasi rectangles. If, however, the pressure at some other depth is required, a new sketch will have to be drawn to such a scale that the distance *AB* on Fig. 3-11 equals the depth at which pressure is desired.

3-4 PROBLEMS

3-1 A concentrated load of 200 kips is applied to the ground surface. What is the vertical stress increment in psf due to the load at a depth of 15 ft directly below the load?

3-2 A concentrated load of 200 kips is applied to the ground surface. What is the vertical stress increment in psf due to the load at a point located at a depth of 15 ft below the ground surface at a horizontal distance of 10 ft from the line of the concentrated load?

3-3 A 10-ft by 7.5-ft rectangular area carrying a uniform load of 5000 psf is applied to the ground surface. Determine the vertical stress increment due to this uniform load at a depth of 12 ft below the ground surface by the approximate method (i.e., 2:1 slope method).

3-4 A circular area carrying a uniform load of 4500 psf is applied to the ground surface. The radius of the circular area is 12 ft. What is the vertical stress increment in psf due to this uniform load (1) at a point 18 ft below the center of the circular area, and (2) at a point 18 ft below the ground surface at a horizontal distance of 6 ft from the center of the circular area?

3-5 An 8-ft by 12-ft rectangular area carrying a uniform load of 6000 psf is applied to the ground surface. What is the vertical stress increment in psf due to the uniform load at a depth of 15 ft below the corner of the rectangular loaded area?

3-6 A 12-ft by 12-ft square area carrying a uniform load of 5000 psf is applied to the ground surface. Find the vertical stress increment in psf due to the load at a depth of 25 ft below the center of the loaded area.

3-7 An L-shaped area shown in Fig. 3-12 carries a 2000-psf uniform load. Find the vertical stress increment in psf due to the structure load at a depth of 24 ft (1) below the interior corner *A*, and (2) below the exterior corner *E*.

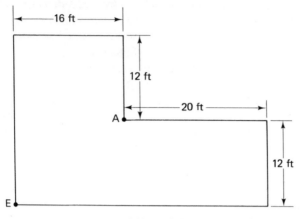

FIGURE 3-12

3-8 A square area *ABCD* shown in Fig. 3-13 carries a 2500-psf uniform load. Find the vertical stress increment due to the exerted load at a depth of 12 ft (1) below point *G*, and (2) below point *J*.

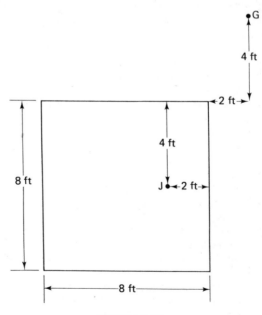

FIGURE 3-13

References

[1] JOSEPH E. BOWLES, *Foundation Analysis and Design*, McGraw-Hill Book Company, New York, 1968.

[2] H. M. WESTERGAARD, "A Problem of Elasticity Suggested by a Problem in Soil Mechanics: Soft Material Reinforced by Numerous Strong Horizontal Sheets," in *Contributions to the Mechanics of Solids*, Stephen Timoshenko 60th Anniversary Volume, Macmillan Publishing Co., Inc., New York, 1938.

[3] MERLIN G. SPANGLER AND RICHARD L. HANDY, *Soil Engineering*, 3rd ed., Intext Educational Publishers, New York, 1973. Copyright 1951, © 1960, 1973 by Harper & Row, Publishers, Inc. Reprinted by permission of the publisher.

[4] NATHAN M. NEWMARK, *Simplified Computation of Vertical Pressures in Elastic Foundations*, Circ. No. 24, Eng. Exp. Sta., Univ. Ill., 1935.

[5] T. WILLIAM LAMB AND ROBERT V. WHITMAN, *Soil Mechanics, SI Version*, John Wiley & Sons, Inc., New York, 1979. Copyright © 1979, by John Wiley & Sons, Inc. Reprinted by permission of John Wiley & Sons, Inc.

[6] NATHAN M. NEWMARK, *Influence Charts for Computation of Stresses in Elastic Foundations*, Univ. Ill. Bull. 338, 1942.

[7] WAYNE C. TENG, *Foundation Design*, Prentice-Hall, Inc., Englewood Cliffs, N.J., 1962.

4

Consolidation of Soil and Settlement of Structures

4-1 INTRODUCTION

Structures built on soil are subject to settlement. Some settlement is often inevitable; and, depending on circumstances, some settlement is tolerable. For example, small uniform settlement of a building throughout the floor area might be tolerable, whereas nonuniform settlement of the same building might not be. Or, settlement of a garage or warehouse building might be tolerable, whereas the same settlement (especially differential settlement) of a luxury apartment building would not be because of damage to walls, ceilings, and so on. In any event, a knowledge of the causes of settlement and a means of computing (or predicting) settlement quantitatively are important to the soils engineer and technologist.

Although there are several possible causes of settlement (e.g., dynamic forces, changes in the groundwater table, adjacent excavation, etc.), probably the major cause is compressive deformation of soil beneath the structure. Such compressive deformation generally results from reduction in void volume. This reduction in void volume is accompanied by a rearrangement of the soil grains and a compression of the material in the voids. If the soil is dry, the voids are filled with air; and since air is compressible, the rearrangement of the soil grains can occur rapidly. If the soil is saturated, the voids are filled with incompressible water, and water must be extruded from the soil mass before the soil grains can rearrange themselves. In soils of high permeability (i.e., coarse-grained soils), this process requires a short time interval for completion. The result is that almost all of the settlement has occurred by the time the construction is complete. However, in soils of low permeability (i.e., fine-grained soils), the process requires a long time interval for completion. The result is that the strain occurs very slowly; thus, settlement will take place slowly and will continue over a long period of time. The latter case (fine-grained soil) is of more concern because of long-term uncertainty.

As indicated, the process of compression due to extrusion of water from the voids in a fine-grained soil as a result of increased loading (such as the weight of a structure above) is very slow and continues over a long period of time. This phenomenon is called *consolidation*. [1] Associated settlement is referred to as *consolidation settlement*. The analysis of consolidation settlement (fine-grained soil) is rather involved, and most of this chapter is required to present it. The analysis of settlement on coarse-grained, cohesionless soil is simpler and is discussed in Sec. 4-7.

In analyzing a clayey soil (fine-grained soil), it is necessary to differentiate between two types of clay—*normally consolidated* clay and *overconsolidated* clay. In the case of normally consolidated clay, the clay formation has never been subjected to any loading larger than the present effective overburden pressure. This is true whenever the height of soil above the clay formation (and therefore the weight of the soil above, which causes the pressure) has been more or less constant through time. In the case of overconsolidated clay, the clay formation has been subjected at some time to a loading greater than the present effective overburden pressure. This is true whenever the present height of soil above the clay formation is less than it was at some time in the past. Such a situation could exist if significant erosion has occurred at the ground surface. (Because of the erosion, the present height of soil above the clay formation is less than it was prior to the erosion.) It might be noted that overconsolidated clay is generally less compressible. As will be related in this chapter, the analysis of clays for settlement differs somewhat depending on whether the clay is normally consolidated or overconsolidated.

This chapter deals primarily with the determination of settlement of structures. Sections 4-2 through 4-6 deal with settlement on clay. Section 4-2 deals with the laboratory testing required for analyzing settlement. Section 4-3 shows how the laboratory data are analyzed to determine if the clay is normally consolidated, and Sec. 4-4 shows how they are analyzed to determine if the clay is overconsolidated. Section 4-5 demonstrates the development of a "field consolidation line," which in turn is used to calculate settlement on clay (Sec. 4-6). Section 4-7 deals with settlement on sand.

4-2 CONSOLIDATION TEST

As a means of estimating both the amount and the time of consolidation and resulting settlement, consolidation tests are run in the laboratory. For complete and detailed instructions for conducting a consolidation test, the reader is referred to a soils laboratory manual. A generalized discussion is given here.

To begin with, an undisturbed soil sample is placed in a metal ring. One porous disk is placed above the sample and another is placed beneath the sample. The purpose of the disks is to allow water to flow into and out of the soil sample. This assembly is immersed in water. As load is applied to the upper disk, the sample is compressed and deformation is measured by a dial gauge (see Fig. 4-1).

To begin a particular test, a specific pressure (e.g., $\frac{1}{8}$ ton/ft²) is applied to the soil sample, and dial readings (reflecting deformation) and corresponding time observations are made and recorded until deformation has nearly ceased. Normally, this is done over a 24-hour period. Using these data, a graph is prepared with time along the abscissa on a logarithmic scale and dial readings along the ordinate on an arithmetic scale. An example of such a graph is given in Fig. 4-2.

The procedure described above is repeated after doubling the applied pressure, giving another graph of time versus dial readings corresponding to this pressure. The procedure is repeated for additional doublings of the applied pressure until the applied pressure is in excess of the total pressure to which the clay formation is expected to be subjected when the proposed structure is built. [The total pressure includes the effective overburden pressure and net additional pressure (or consolidation pressure) due to the structure.]

From each graph of time versus dial readings, it is possible to determine the void ratio (e) and the coefficient of consolidation (C_v) that correspond to the specific applied pressure or loading (p) for that graph. Using these data, two graphs can be prepared—one of void ratio versus pressure (e–$\log p$ curve), with pressure along the abscissa on a logarithmic scale and void ratio along the ordinate on an arithmetic scale, and another of consolidation coefficient versus pressure (C_v–$\log p$ curve), with pressure along the abscissa on a logarithmic scale and coefficient of consolidation along the ordinate on an arithmetic scale. An example of an e–$\log p$ curve is given in Fig. 4-3, and an example of a C_v–$\log p$ curve is given in Fig. 4-4. As will be related subsequently, the e–$\log p$ curve is used to determine the amount of settlement, and the C_v–$\log p$ curve is used to determine the timing of the settlement.

In Fig. 4-3, the upper curve exhibits the relationship between void ratio and pressure as the pressure is increased. As will be shown in Sec. 4-5, in the case of overconsolidated clay, it is necessary to have a "rebound curve."

Exhibited by the lower curve in Fig. 4-3, the rebound curve is obtained by unloading the soil sample during the consolidation test after the maximum pressure has been reached. As the sample is unloaded, the soil tends to swell, causing movement and associated dial readings to reverse direction.

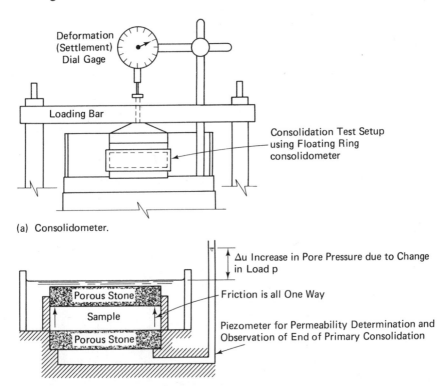

(a) Consolidometer.

(b) Fixed-Ring Consolidometer. May be used to Obtain Permeability Information during a Consolidation Test if a Piezometer is Installed.

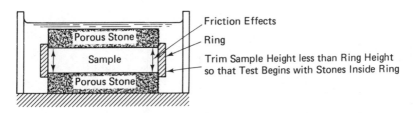

(c) Floating Ring Consolidometer.

FIGURE 4-1 (a) Consolidometer. (b) Fixed-ring consolidometer. May be used to obtain permeability information during a consolidation test if a piezometer is installed. (c) Floating-ring consolidometer. [2]

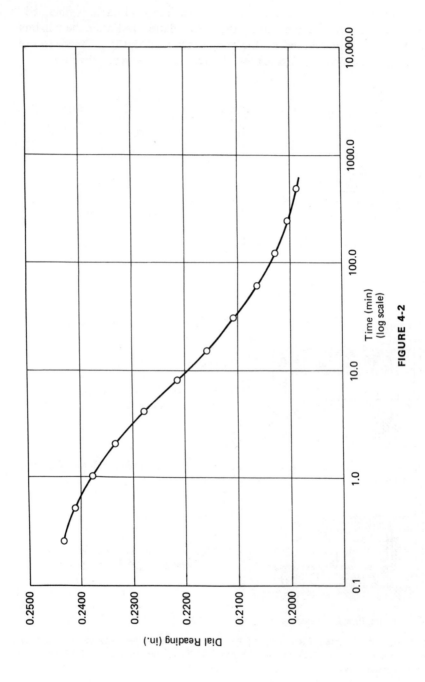

Dial Reading (in.)

Time (min)
(log scale)

FIGURE 4-2

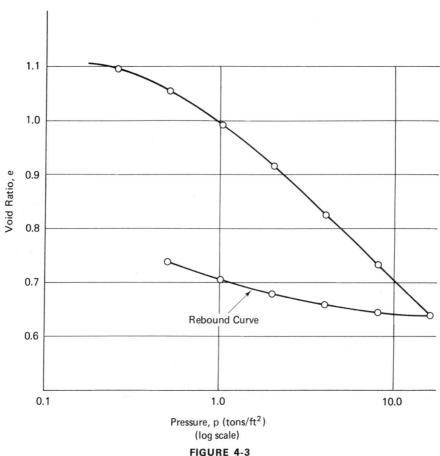

FIGURE 4-3

The primary results of a laboratory consolidation test are (1) the *e*–log *p* curve, (2) the C_v–log *p* curve, and (3) the initial void ratio of the soil *in situ* (e_0).

4-3 NORMALLY CONSOLIDATED CLAY

As indicated in Sec. 4-1, in the case of a clay soil it is necessary to determine whether the clay is either normally consolidated or overconsolidated. This section shows how to determine if a given clay soil is normally consolidated.

It is first necessary, however, to determine the present effective overburden pressure (P_0). This pressure is the result of the (effective) weight of soil above midheight of the consolidating clay layer. Although the reader

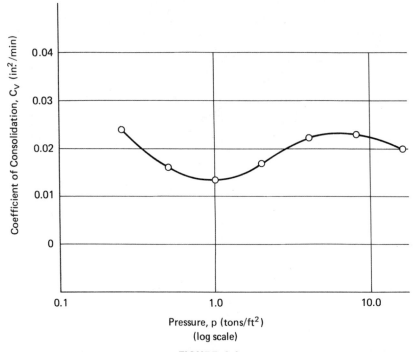

FIGURE 4-4

probably knows how to calculate effective overburden pressure, the procedure is illustrated in Example 4-1.

EXAMPLE 4-1

Given

The profile of the soil is shown in Fig. 4-5.

Required

Determine the present effective overburden pressure (P_0) at the midheight of the compressible clay layer.

Solution

Elevation of the midheight of the clay layer $= \dfrac{732 + 710}{2} = 721$

Overburden pressure (P_0) at the midheight of the compressible clay layer
$= (132)(760-752) + (132-62.4)(752-732) + (125.4-62.4)(732-721)$
$= 3141 \text{ psf} = 1.57 \text{ tons/ft}^2$

Elev. 760 ft

Sand and Gravel
  Unit Weight = 132.0 pcf   ▽   Water Table   Elev. 752 ft

Sand and Gravel

    Unit Weight = 132.0 pcf

Elev. 732 ft

Clay

    Unit Weight = 125.4 pcf

Elev. 710 ft

FIGURE 4-5

The first step in determining if a given clay soil is normally consolidated is to locate the point designated by a pressure of P_0 (distance along the abscissa) and a void ratio of e_0 (distance along the ordinate). This point is labeled *a* in Fig. 4-6. The next step is to project the lower right straight-line portion of the *e–log p* curve in a straight line upward and to the left. This is the dashed line in Fig. 4-6; it will intersect a horizontal line drawn at *e* equal to e_0. The point of intersection of these two lines is labeled *b* in Fig. 4-6. If point *b* is to the left of point *a* (as in Fig. 4-6), the soil is normally consolidated clay [3].

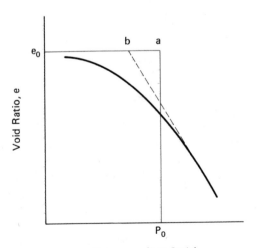

FIGURE 4-6 [3]

4-4 OVERCONSOLIDATED CLAY

The procedure for determining if a given clay is overconsolidated clay is essentially the same as that for determining if the sample is normally consolidated clay. The point designated by a pressure of P_0 and a void ratio of e_0 is located and labeled a. The lower right portion of the e–$\log p$ curve is projected in a straight line upward and to the left until it intersects a horizontal line drawn at e equal to e_0, with the point of intersection labeled b. If point b is to the right of point a (as in Fig. 4-7), the soil is overconsolidated clay [3].

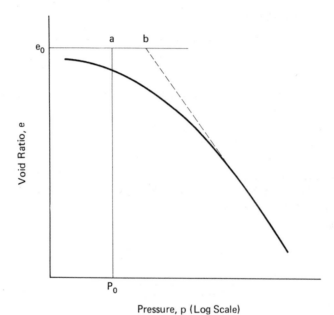

FIGURE 4-7 [3]

If the given clay is believed to be overconsolidated clay, it is necessary to determine (for subsequent analysis of settlement) the maximum overburden pressure at the consolidated clay layer (P_0'). The following procedure, developed by Casagrande [4], can be used to determine P_0'. The first step is to locate the point on the e–$\log p$ curve where the curvature is greatest (where the radius of curvature is smallest). This is indicated by point g in Fig. 4-8. From this point two straight lines are drawn—one horizontal line (line gh in Fig. 4-8) and one line tangent to the e–$\log p$ curve (line gj in Fig. 4-8). The next step is to draw a line that bisects the angle between lines gh and gj (line gi in Fig. 4-8). The final step is to project the lower right straight-line portion of the e–$\log p$ curve in a straight line upward and to the left. This

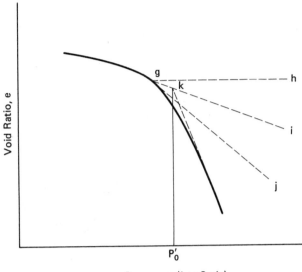

Pressure, p (Log Scale)

FIGURE 4-8   [3, 4]

projected line will intersect line gi at a point such as k in Fig. 4-8. The value of p corresponding to point k (p coordinate of point k along the abscissa) is taken as P'_0 [3].

4-5 FIELD CONSOLIDATION LINE

The e–$\log p$ curves considered in previous sections give, of course, the relationship between void ratio and pressure for a given soil. Such a relationship is used in calculating settlement. The e–$\log p$ curves of Fig. 4-3 and Figs. 4-6 through 4-8 reflect, however, the relationship between void ratio and pressure for the soil sample in the laboratory. Although an "undisturbed sample" is used in the laboratory test, it is not generally possible to duplicate the soil in the laboratory exactly as it exists in the field. Thus, the e–$\log p$ curves developed from laboratory consolidation tests are modified to give an e–$\log p$ curve that is presumed to reflect actual field conditions. This modified e–$\log p$ curve is called the *field consolidation line.* Two methods for determining the field consolidation line follow—one for normally consolidated clay and one for overconsolidated clay.

In the case of normally consolidated clay, the determination of the field consolidation line is fairly simple. With the given e–$\log p$ curve developed from the laboratory test (Fig. 4-9), the point on the e–$\log p$ curve corresponding to $0.4e_0$ is determined (point f in Fig. 4-9). A straight line connecting points

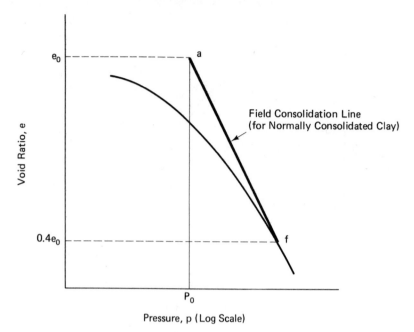

FIGURE 4-9 [3]

a and f gives the field consolidation line for the normally consolidated clay
[3, 5]. (The reader will recall that, as related in Sec. 4-3 and Fig. 4-6, point a
is the point designated by a pressure of P_0 and a void ratio of e_0.)

In the case of overconsolidated clay, the determination of the field consolidation line is somewhat more difficult. With the given e–$\log p$ curve developed from the laboratory test (Fig. 4-10), the point on the e–$\log p$ curve
corresponding to $0.4e_0$ is determined (point f in Fig. 4-10). Point a (the point
designated by a pressure of P_0 and a void ratio of e_0) is located, and a line is
drawn through point a parallel to the rebound line. This line through point
a parallel to the rebound line is shown as a dashed line in Fig. 4-10; it will
intersect a vertical line drawn at $p = P_0'$. (The procedure for evaluating P_0'
was given in Sec. 4-4.) This point of intersection is designated by m in Fig.
4-10. Points m and f are connected by a straight line, and points a and m are
connected by a curved line that follows the same general shape of the e–$\log p$
curve. This curved line from a to m and the straight line from m to f give the
field consolidation line for overconsolidated clay [6, 7].

It is the field consolidation line—the dark line in Fig. 4-9 (normally consolidated clay) and the dark line in Fig. 4-10 (overconsolidated clay)—that is
used in calculating settlement. The other curves (dial readings versus time and
e–$\log p$ curve) are required only as a means of determining the field consolida-

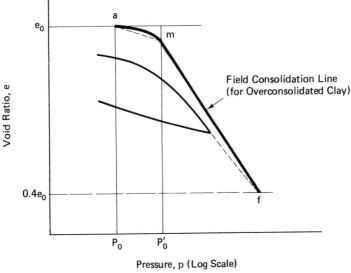

FIGURE 4-10 [6, 7]

tion line. Once the field consolidation line is determined, these other curves are no longer used in determining settlement.

Of special significance is the *slope* of the field consolidation line. This slope is called the *compression index* (C_c). It may be evaluated by finding the coordinates of any two points on the field consolidation line [(p_1, e_1) and (p_2, e_2)] and substituting these values into the equation [3]

$$C_c = \frac{e_1 - e_2}{\log p_2 - \log p_1} = \frac{e_1 - e_2}{\log (p_2/p_1)} \qquad (4\text{-}1)$$

Skempton has shown that the compression index can be approximated in terms of the liquid limit (*LL*, in percent) by the equation [3, 8]

$$C_c = 0.009(LL - 10) \qquad (4\text{-}2)$$

for normally consolidated clays.

It should be emphasized that the value of C_c computed from Eq. (4-1) is obtained from the field consolidation line, which is based on the results of a consolidation test, while that computed from Eq. (4-2) is based solely on the liquid limit. The consolidation test is much more lengthy, difficult, and expensive to perform than the test to determine the liquid limit. Also, the calculation of C_c using results of a consolidation test is much more involved than the calculation using the liquid limit. However, the calculation of C_c

using the liquid limit [Eq. (4-2)] is only an approximation and should be used only when very approximate values of settlement are acceptable (such as in a preliminary design).

EXAMPLE 4-2

Given

A normally consolidated clay has a liquid limit of 51.2%.

Required

Estimate the compression index (C_c).

Solution

From Eq. (4-2),

$$C_c = 0.009(LL - 10) \qquad (4\text{-}2)$$
$$= 0.009(51.2 - 10)$$
$$= 0.37$$

4-6 SETTLEMENT OF LOADS ON CLAY

Once the field consolidation line is determined for a given case, the total expected consolidation settlement of load on clay can be computed from the equation [1, 6]

$$S = \frac{e_0 - e}{1 + e_0}[H] \qquad (4\text{-}3)$$

where S = total settlement due to consolidation, in.

e_0 = initial void ratio of the soil *in situ*

e = void ratio of the soil corresponding to the total pressure (p) acting on the midheight of the consolidating clay layer

H = thickness of the clay layer, in.

In practice, the value of e_0 is obtained from the laboratory consolidation test, and the value of e is obtained from the field consolidation line based on the total pressure (i.e., effective overburden pressure plus net additional pressure due to structure—both at midheight of the consolidating clay layer). The value of H is obtained from soil exploration (Chap. 2).

An alternative mathematical means of computing the total expected consolidation settlement uses the slope of the field consolidation line (C_c) and the equation [1, 6]

$$S = C_c\left(\frac{H}{1 + e_0}\right)\log\frac{p}{P_0} \qquad (4\text{-}4)$$

where C_c = slope of the field consolidation line (compression index)

p = total pressure acting on the midheight of the consolidating clay layer = $P_0 + \Delta p$

P_0 = present effective overburden pressure at midheight of the consolidating clay layer

Δp = net additional pressure at midheight of the consolidating clay layer due to structure

The value of C_c can be determined by evaluating the slope of the field consolidation line [Eq. (4-1)] or approximated based on the liquid limit [Eq. (4-2)]. If the latter method is used, the computed settlement should be considered as a rough approximation.

The time rate of settlement due to consolidation can be computed from the equation [1, 6]

$$t = \frac{T_v}{C_v} H^2 \tag{4-5}$$

where t = time to reach a particular percent of consolidation; the percent of consolidation is defined as the ratio of the amount of settlement at a certain time during the process of consolidation to the total settlement due to consolidation

T_v = time factor, a coefficient depending on the particular percent of consolidation

C_v = coefficient of consolidation corresponding to the total pressure ($p = P_0 + \Delta p$) acting on the midheight of the clay layer

H = thickness of the consolidating clay layer. [However, if the clay layer *in situ* is drained on both top and bottom, half the thickness of the layer should be substituted for H in Eq. (4-5)]

In practice, the value of T_v is determined from Fig. 4-11, based on the desired percent of consolidation (U), and the value of C_v is determined from the C_v–log p curve (e.g., Fig. 4-4) based on the total pressure acting on the midheight of the clay layer. It will be recalled that the C_v–log p curve is a product of the laboratory consolidation test.

In summary, either Eq. (4-3) or Eq. (4-4) can be used to compute total settlement due to consolidation. Then Eq. (4-5) can be used to compute the time required to reach a particular percentage of that consolidation. For example, if the total settlement due to consolidation is computed to be 3.0 in., the time required for the structure to settle 1.5 in. could be computed from Eq. (4-5) by substituting a value of T_v of 0.196 (along with the applicable

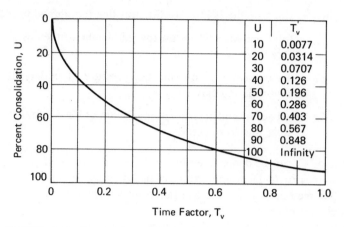

U	T_v
10	0.0077
20	0.0314
30	0.0707
40	0.126
50	0.196
60	0.286
70	0.403
80	0.567
90	0.848
100	Infinity

FIGURE 4-11 Time factor as a function of percentage of consolidation. [1]

values of C_v and H). The value of 0.196 is obtained from Fig. 4-11 for a value of U of 50%. U is 50% because the particular settlement being considered (1.5 in.) is 50% of the total settlement (3.0 in.).

Examples 4-3 through 4-5 should assist the reader's understanding of the calculation of the amount and time rate of consolidation settlement.

EXAMPLE 4-3

Given

1. A sample of normally consolidated clay was obtained by a Shelby tube sampler from the midheight of a compressible clay layer (see Fig. 4-12).

2. A consolidation test was conducted on a portion of this sample. The results of the consolidation test are as follows:
    a. Natural (or initial) void ratio of the clay existing in the field (e_0) is 1.65.
    b. The pressure–void ratio relations are as follows:

p (tons/ft^2)	e
0.8	1.50
1.6	1.42
3.2	1.30
6.4	1.12
12.8	0.94

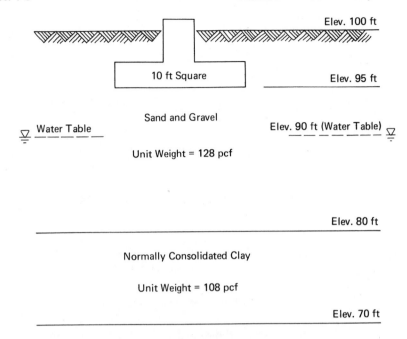

Elev. 100 ft

10 ft Square

Elev. 95 ft

Sand and Gravel

Water Table

Elev. 90 ft (Water Table)

Unit Weight = 128 pcf

Elev. 80 ft

Normally Consolidated Clay

Unit Weight = 108 pcf

Elev. 70 ft

FIGURE 4-12

3. A footing is to be located 5 ft below the ground level, as shown in Fig. 4-12. The base of the square footing is 10 ft by 10 ft and it exerts a total load of 250 tons, which includes column load, weight of footing, and weight of soil surcharge on the footing.

Required

1. From the results of the consolidation test given above, prepare an *e*–**log** *p* curve and construct a field consolidation line, assuming that point *f* is located at $0.4e_0$.

2. Compute the total expected settlement for the clay layer.

Solution

1. Present effective overburden pressure (P_0) at midheight of clay layer

$$= (128)(100 - 90) + (128 - 62.4)(90 - 80)$$
$$+ (108 - 62.4)\left(\frac{80 - 70}{2}\right)$$

$P_0 = 2164 \text{ psf} = 1.08 \text{ tons/ft}^2$

$e_0 = 1.65 \text{ (given)}$

$0.4e_0 = (0.4)(1.65) = 0.66$

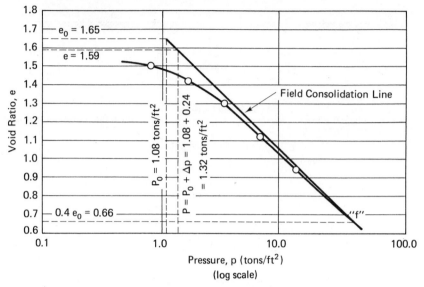

FIGURE 4-13 *e–log p* curve.

The *e*–log *p* curve is shown in Fig. 4-13 together with the field consolidation line.

2. The effective weight of excavation $= (128)(5) = 640$ psf $= 0.32$ ton/ft². Net consolidation pressure at the base of footing

$$= \frac{250}{10 \times 10} - 0.32 = 2.18 \text{ tons/ft}^2$$

To determine the net consolidation pressure at midheight of the clay layer under the center of the footing, it is necessary to divide the base of the footing into four equal 5-ft by 5-ft square areas. Since each of these square areas has a common corner at the center of the footing, the desired net consolidation pressure at midheight of the clay layer can be calculated upon determining an influence coefficient using either Table 3-2 or Fig. 3-5. Referring to Fig. 3-5,

$$mz = 5 \text{ ft}; \quad z = 95 - \frac{80 + 70}{2} = 20 \text{ ft}; \quad m = \frac{5}{20} = 0.25$$

$$nz = 5 \text{ ft}; \quad z = 20 \text{ ft}; \quad n = \frac{5}{20} = 0.25$$

From Fig. 3-5, the influence coefficient $= 0.027$. The net consolidation pressure at midheight of the clay layer under the center of the footing

$$\Delta p = (4)(0.027)(2.18) = 0.24 \text{ ton/ft}^2$$

Final pressure at midheight of clay layer

$$p = P_0 + \Delta p = 1.08 + 0.24 = 1.32 \text{ tons/ft}^2$$

Enter $p = 1.32$ tons/ft^2 along the abscissa of the e–$\log p$ curve (Fig. 4-13) and move upward vertically until the "field consolidation line" is intersected. Then turn left and move horizontally to read a void ratio e of 1.59 on the ordinate of the e–$\log p$ curve. With

$$e_0 = 1.65$$
$$e = 1.59$$
$$H = 10 \text{ ft} = 120 \text{ in.}$$

substitute into Eq. (4-3):

$$S = \frac{e_0 - e}{1 + e_0}[H] \qquad\qquad (4\text{-}3)$$

$$= \frac{1.65 - 1.59}{1 + 1.65}[120] = 2.72 \text{ in.}$$

The total expected settlement is 2.72 in.

EXAMPLE 4-4

Given

1. Same as Example 4-3; total settlement = 2.72 in.

2. The results of the consolidation test conducted in the soil laboratory also indicated that the coefficient of consolidation for the clay sample (C_v) is 3.28×10^{-3} in.^{2}/min for the pressure increment from 0.8 to 1.6 tons/ft^2.

Required

Compute the time of settlement (years) assuming that:

(a) The clay layer is underlain by permeable sand and gravel (double drainage).

(b) The clay layer is underlain by impermeable bedrock (single drainage).

Take U at 10% increments and plot these values on a **settlement–log time curve**.

Solution

(a) Clay layer is underlain by permeable sand and gravel (double drainage).

Use Eq. (4-5):

$$t = \frac{T_v}{C_v} H^2 \tag{4-5}$$

where $C_v = 3.28 \times 10^{-3}$ in.^{2}/min

$$H = \frac{10}{2}\,\text{ft} = 5\,\text{ft} \quad \text{(double drainage)}$$

(1) When $U = 10\%$ (i.e., 10% of total settlement, $S_{10} = 2.72 \times 10\% = 0.27$ in.),

$$T_v = 0.0077 \quad \text{(from Fig. 4-11)}$$

$$t_{10} = \frac{(0.0077)(5 \times 12)^2 \text{ in.}^2}{3.28 \times 10^{-3} \text{ in.}^2/\text{min}} = 8451 \text{ min} = 0.016 \text{ yr}$$

This indicates that the footing will settle approximately 0.27 in. in 0.016 yr.

(2) When $U = 20\%$ (i.e., 20% of total settlement, $S_{20} = 0.54$ in.),

$$T_v = 0.0314 \quad \text{(from Fig. 4-11)}$$

$$t_{20} = \frac{(0.0314)(5 \times 12)^2}{3.28 \times 10^{-3}} = 34{,}463 \text{ min} = 0.066 \text{ yr}$$

This indicates that the footing will settle approximately 0.54 in. in 0.066 yr.

(3) When $U = 30\%$ (i.e., 30% of total settlement),

$$T_v = 0.0707 \quad \text{(from Fig. 4-11)}$$

$$t_{30} = \frac{(0.0707)(5 \times 12)^2}{3.28 \times 10^{-3}} = 77{,}598 \text{ min} = 0.15 \text{ yr}$$

(4) When $U = 40\%$ (i.e., 40% of total settlement),

$$T_v = 0.126 \quad \text{(from Fig. 4-11)}$$

$$t_{40} = \frac{(0.126)(5 \times 12)^2}{3.28 \times 10^{-3}} = 1.383 \times 10^5 \text{ min} = 0.26 \text{ yr}$$

(5) When $U = 50\%$ (i.e., 50% of total settlement),

$$T_v = 0.196 \quad \text{(from Fig. 4-11)}$$

$$t_{50} = \frac{(0.196)(5 \times 12)^2}{3.28 \times 10^{-3}} = 2.151 \times 10^5 \text{ min} = 0.41 \text{ yr}$$

(6) When $U = 60\%$ (i.e., 60% of total settlement),

$$T_v = 0.286 \quad \text{(from Fig. 4-11)}$$

$$t_{60} = \frac{(0.286)(5 \times 12)^2}{3.28 \times 10^{-3}} = 3.139 \times 10^5 \text{ min} = 0.60 \text{ yr}$$

(7) When $U = 70\%$ (i.e., 70% of total settlement),

$$T_v = 0.403 \quad \text{(from Fig. 4-11)}$$

$$t_{70} = \frac{(0.403)(5 \times 12)^2}{3.28 \times 10^{-3}} = 4.423 \times 10^5 \text{ min} = 0.84 \text{ yr}$$

(8) When $U = 80\%$ (i.e., 80% of total settlement),

$$T_v = 0.567 \quad \text{(from Fig. 4-11)}$$

$$t_{80} = \frac{(0.567)(5 \times 12)^2}{3.28 \times 10^{-3}} = 6.223 \times 10^5 \text{ min} = 1.18 \text{ yr}$$

(9) When $U = 90\%$ (i.e., 90% of total settlement),

$$T_v = 0.848 \quad \text{(from Fig. 4-11)}$$

$$t_{90} = \frac{(0.848)(5 \times 12)^2}{3.28 \times 10^{-3}} = 9.307 \times 10^5 \text{ min} = 1.77 \text{ yr}$$

(b) Clay layer is underlain by impermeable bedrock (single drainage). Eq. (4-5) is still applicable.

$$t = \frac{T_v}{C_v} H^2 \tag{4-5}$$

where $C_v = 3.28 \times 10^{-3}$ in.^{2}/min

$\qquad\quad H = 10$ ft   (single drainage)

(1) When $U = 10\%$,

$$T_v = 0.0077$$

$$t_{10} = \frac{(0.0077)(10 \times 12)^2}{3.28 \times 10^{-3}} = 33,805 \text{ min} = 0.064 \text{ yr}$$

(2) When $U = 20\%$,

$$T_v = 0.0314$$

$$t_{20} = \frac{(0.0314)(10 \times 12)^2}{3.28 \times 10^{-3}} = 1.379 \times 10^5 \text{ min} = 0.26 \text{ yr}$$

(3) When $U = 30\%$,

$$T_v = 0.0707$$

$$t_{30} = \frac{(0.0707)(10 \times 12)^2}{3.28 \times 10^{-3}} = 3.104 \times 10^5 \text{ min} = 0.59 \text{ yr}$$

(4) When $U = 40\%$,

$$T_v = 0.126$$

$$t_{40} = \frac{(0.126)(10 \times 12)^2}{3.28 \times 10^{-3}} = 5.532 \times 10^5 \text{ min} = 1.05 \text{ yr}$$

(5) When $U = 50\%$,

$$T_v = 0.196$$

$$t_{50} = \frac{(0.196)(10 \times 12)^2}{3.28 \times 10^{-3}} = 8.605 \times 10^5 \text{ min} = 1.64 \text{ yr}$$

(6) When $U = 60\%$,

$$T_v = 0.286$$

$$t_{60} = \frac{(0.286)(10 \times 12)^2}{3.28 \times 10^{-3}} = 1.256 \times 10^6 \text{ min} = 2.39 \text{ yr}$$

(7) When $U = 70\%$,

$$T_v = 0.403$$

$$t_{70} = \frac{(0.403)(10 \times 12)^2}{3.28 \times 10^{-3}} = 1.769 \times 10^6 \text{ min} = 3.37 \text{ yr}$$

(8) When $U = 80\%$,

$$T_v = 0.567$$

$$t_{80} = \frac{(0.567)(10 \times 12)^2}{3.28 \times 10^{-3}} = 2.489 \times 10^6 \text{ min} = 4.74 \text{ yr}$$

(9) When $U = 90\%$,

$$T_v = 0.848$$

$$t_{90} = \frac{(0.848)(10 \times 12)^2}{3.28 \times 10^{-3}} = 3.723 \times 10^6 \text{ min} = 7.08 \text{ yr}$$

The results of these computations are tabulated in Table 4-1 and are shown graphically by a **settlement–log time** curve in Fig. 4-14.

TABLE 4-1   Computed time–settlement relation.

Fraction of Total Settlement, U (%)	Settlement (in.)	TIME (yr) Double Drainage	Single Drainage
10	0.27	0.016	0.064
20	0.54	0.066	0.26
30	0.82	0.15	0.59
40	1.09	0.26	1.05
50	1.36	0.41	1.64
60	1.63	0.60	2.39
70	1.90	0.84	3.37
80	2.18	1.18	4.74
90	2.45	1.77	7.08
100	2.72	∞	∞

EXAMPLE 4-5

Given

1. An 8-ft clay layer beneath a building is overlain by a stratum of permeable sand and gravel and is underlain by impermeable bedrock.
2. The total expected settlement for the clay layer due to the footing load is 2.5 in.
3. The coefficient of consolidation (C_v) is 2.68×10^{-3} in.²/min.

Required

1. How many years will it take for 90% of the total expected settlement to take place?
2. Compute the amount of settlement that will occur in 1 year.
3. How many years will it take for 1 in. of settlement to take place?

Solution

1. From Eq. (4-5),

$$t = \frac{T_v}{C_v} H^2 \tag{4-5}$$

$T_v = 0.848$ (for $U = 90\%$; see Fig. 4-11)

$C_v = 2.68 \times 10^{-3}$ in.²/min

$H = 8$ ft   (single drainage)

$$t_{90} = \frac{(0.848)(8 \times 12)^2}{2.68 \times 10^{-3}} = 2.916 \times 10^6 \text{ min} = 5.55 \text{ yr}$$

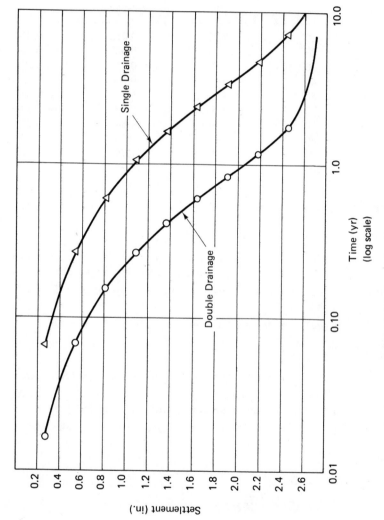

FIGURE 4-14 Settlement–log time curve.

2. From Eq. (4-5),

$$t = \frac{T_v}{C_v} H^2 \tag{4-5}$$

$t = 1 \text{ yr}$

$C_v = 2.68 \times 10^{-3} \text{ in.}^2/\text{min}$

$H = 8 \text{ ft}$

$$1 = \frac{T_v}{2.68 \times 10^{-3}} (8 \times 12)^2 \times \frac{1}{(60)(24)(365)}$$

$$T_v = \frac{(2.68 \times 10^{-3})(60)(24)(365)}{(8 \times 12)^2} = 0.15$$

From Fig. 4-11, with $T_v = 0.15$, $U = 43\%$.
The amount of settlement that will occur in 1 yr

$= \text{total settlement multiplied by } U\% = (2.5)(0.43) = 1.1 \text{ in.}$

3. $U\% = \text{fraction of total settlement}$

$$U = \frac{1 \text{ (in.)}}{\text{total settlement (in.)}} = \frac{1}{2.5} \times 100 = 40\%$$

From Fig. 4-11, with $U = 40\%$, $T_v = 0.126$.
From Eq. (4-5),

$$t = \frac{T_v}{C_v} H^2 \tag{4-5}$$

$$= \frac{(0.126)(8 \times 12)^2}{2.68 \times 10^{-3}} = 4.333 \times 10^5 \text{ min} = 0.82 \text{ yr}$$

4-7 SETTLEMENT OF LOADS ON SAND

Most of the settlement of loads on sand has occurred by the time the construction is complete. Thus time rate of settlement is not a factor as it is in the case of clay. Neither is the calculation of settlement on sand amenable to solution based on laboratory consolidation test. Instead, calculation of settlement on sand is generally done by empirical means.

One such empirical means is based on the standard penetration test (SPT), which was discussed in Sec. 2-4. To determine settlement on sand, SPT determinations should be made at various depths at the test site. These SPT determinations are normally made at depth intervals of $2\frac{1}{2}$ ft, beginning at the depth corresponding to the base of the proposed footing. These SPT values must be corrected for overburden pressure (see Chap. 2). The next step is to compute the average corrected SPT value for each boring for the sand between the base of the footing and a depth B below the base of the footing, where B is the width of footing. The lowest of these average corrected SPT values for all the borings at the site is noted and designated N_{lowest}. The maximum settlement can then be computed from the equation [9]

$$s_{max} = \frac{2q}{N_{lowest}} \left[\frac{2B}{1+B}\right]^2 \tag{4-6}$$

where s_{max} = maximum settlement on dry sand, in.

q = applied pressure, tons/ft^2

B = width of footing, ft

Equation (4-6) is applicable to settlement on dry sand. If the groundwater table is located at a depth below the base of the footing less than half the width of the footing, the settlement computed from Eq. (4-6) should be corrected by multiplying it by x_B, where [9]

$$x_B = \frac{P_d}{P_w} \tag{4-7}$$

where P_d = effective overburden pressure at a depth $B/2$ below the base of the footing assuming that the groundwater table is not present

P_w = effective overburden pressure at the same depth with the groundwater table present

Examples 4-6 through 4-9 demonstrate the calculation of settlement on sand.

EXAMPLE 4-6

Given

1. A 10-ft by 10-ft footing carrying a total load of 280 tons is to be constructed on sand as shown in Fig. 4-15.

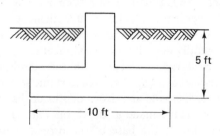

5 ft

10 ft

Medium to Coarse Sand

Unit Weight = 124 pcf

No Groundwater was Encountered

FIGURE 4-15

2. Standard penetration tests were conducted on the site. The results of the tests were corrected for overburden pressure (see Chap. 2), and the corrected SPT values are listed below.

Depth (ft)	Corrected SPT Values, $N_{corrected}$
5.0	31
7.5	36
10.0	30
12.5	28
15.0	35
17.5	33
20.0	31

Required

Estimate the maximum expected settlement for the footing.

Solution

Calculation of the average corrected N-values

The average corrected N-value is determined for each boring for the soil that is located between the level of the base of the footing and a depth B below this level, where B is the width of the footing. In this example, the appropriate depths for calculating the average corrected N-values are 5 to 15 ft. The average corrected N-value is a cumulative average down to the depth indicated.

For a depth of 5 ft,

$$\text{Average corrected } N\text{-value} = 31$$

For a depth of 7.5 ft,

$$\text{Average corrected } N\text{-value} = \frac{31 + 36}{2} = 33$$

For a depth of 10.0 ft,

$$\text{Average corrected } N\text{-value} = \frac{31 + 36 + 30}{3} = 32$$

For a depth of 12.5 ft,

$$\text{Average corrected } N\text{-value} = \frac{31 + 36 + 30 + 28}{4} = 31$$

For a depth of 15.0 ft,

$$\text{Average corrected } N\text{-value} = \frac{31 + 36 + 30 + 28 + 35}{5} = 32$$

The lowest average corrected N-value for design

Subsurface soil conditions generally vary somewhat at most construction sites. The N-value selected for design is usually the lowest average corrected N-value, which in this example is 31 (at depth 12.5 ft).
From Eq. (4-6),

$$s_{\max} = \frac{2q}{N_{\text{lowest}}} \left[\frac{2B}{1+B} \right]^2 \tag{4-6}$$

$$q = \frac{280}{10 \times 10} = 2.8 \text{ tons/ft}^2$$

$$N_{\text{lowest}} = 31$$

$$B = 10 \text{ ft}$$

$$s_{\max} = \frac{(2)(2.8)}{31} \left[\frac{2 \times 10}{1+10} \right]^2 = 0.6 \text{ in. on dry sand}$$

EXAMPLE 4-7

Given

Same conditions as in Example 4-6 except that the groundwater table is located 7 ft below the ground level (see Fig. 4-16).

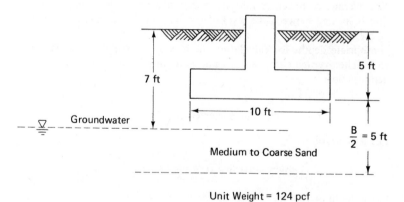

Unit Weight = 124 pcf

FIGURE 4-16

Required

Estimate the maximum expected settlement of the footing.

Solution

From Example 4-6,

$$s_{\max} = 0.6 \text{ in. on dry sand}$$

From Eq. (4-7),

$$x_B = \frac{P_d}{P_w} \tag{4-7}$$

$$P_d = (124)\left(5 + \frac{10}{2}\right) = 1240 \text{ psf}$$

$$P_w = (124)(7) + (124 - 62.4)\left(5 + \frac{10}{2} - 7\right) = 1053 \text{ psf}$$

$$x_B = \frac{P_d}{P_w} = \frac{1240}{1053} = 1.178$$

$$s_{max} = (0.6)(1.178) = 0.7 \text{ in. on wet sand}$$

EXAMPLE 4-8

Given

1. A square footing 8 ft by 8 ft located 5 ft below ground level is to be constructed on sand.

2. Standard penetration tests were conducted on the site. The results of the tests were corrected for overburden pressures, and the lowest average corrected N-value was determined to be 41.

3. Groundwater was not encountered.

Required

The allowable soil pressure for a maximum of 1 in. of settlement.

Solution

From Eq. (4-6),

$$s_{max} = \frac{2q}{N_{lowest}}\left[\frac{2B}{1 + B}\right]^2 \tag{4-6}$$

$$s_{max} = 1 \text{ in.}$$

$$N_{lowest} = 41$$

$$B = 8 \text{ ft}$$

$$1 = \frac{2q}{41}\left[\frac{2 \times 8}{1 + 8}\right]^2$$

$$q = 6.49 \text{ tons/ft}^2 \text{ for 1 in. of settlement}$$

EXAMPLE 4-9

Given

Same conditions as in Example 4-8 except that the groundwater table is located 6 ft below the ground level and the unit weight of the sand is 128 pcf (see Fig. 4-17).

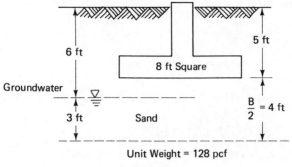

FIGURE 4-17

Required

The allowable soil pressure for a maximum of 1 in. of settlement.

Solution

From Eq. (4-7),

$$x_B = \frac{P_d}{P_w} \qquad (4\text{-}7)$$

$$P_d = (128)\left(5 + \frac{8}{2}\right) = 1152 \text{ psf}$$

$$P_w = (128)(6) + (128 - 62.4)(3) = 964.8 \text{ psf}$$

$$x_B = \frac{1152}{964.8} = 1.194$$

From Example 4-8, allowable soil pressure (q) is 6.49 tons/ft² for 1 in. of settlement when no groundwater is encountered. When the groundwater table is at a depth below the base of the footing less than $B/2$, the s_{max} computed from Eq. (4-6) should be multiplied by x_B. Therefore, in this example an allowable soil pressure (q) of 6.49 tons/ft² will produce $1 \times 1.194 = 1.194$ in. of settlement. Since the settlement varies directly as the bearing pressure,

$$\frac{6.49 \text{ tons/ft}^2}{1.194 \text{ in.}} = \frac{\text{allowable soil pressure for 1 in. of settlement}}{1 \text{ in.}}$$

Allowable soil bearing pressure for 1 in. of settlement = 5.44 tons/ft².

4-8 PROBLEMS

4-1 A sample of normally consolidated clay was obtained by a Shelby tube sampler from the midheight of a compressible clay layer (see Fig. 4-18). A consolidation test was conducted on a portion of this sample, the results of which are

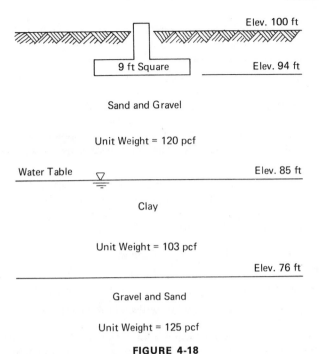

Elev. 100 ft

9 ft Square   Elev. 94 ft

Sand and Gravel

Unit Weight = 120 pcf

Water Table   Elev. 85 ft

Clay

Unit Weight = 103 pcf

Elev. 76 ft

Gravel and Sand

Unit Weight = 125 pcf

FIGURE 4-18

given below:

1. Natural (initial) void ratio of the clay existing in the field $(e_0) = 1.80$.

2. The pressure–void ratio relationships are as follows:

p (tons/ft^2)	e
0.250	1.72
0.500	1.70
1.00	1.64
2.00	1.51
4.00	1.34
8.00	1.15
16.00	0.95

A footing is to be constructed 6 ft below the ground surface, as shown in Fig. 4-18 in the given soil profile. The base of the footing is 9 ft by 9 ft, and it carries a total load of 200 tons, which includes the column load, the weight of the footing, and the weight of soil surcharge on the footing.

(a) From the consolidation test results, prepare an e–$\log p$ curve on three-cycle semilog paper and construct a field consolidation line assuming that the point f is located at $0.4e_0$.

(b) Compute the total expected settlement for the compressible clay layer.

4-2 Continuing Problem 4-1, the results of the consolidation test also indicated that the coefficient of consolidation (C_v) for the clay sample tested is 2.18×10^{-3} in.^{2}/min for the pressure increment from 1 to 2 tons/ft^2. Compute the time of settlement (years). Take U at 10% increments and plot these values on a **settlement–log time** curve.

4-3 A compressible 12-ft clay layer beneath a building is overlain by a stratum of sand and gravel and is underlain by impermeable bedrock. The total expected settlement for the compressible clay layer due to the building load is 4.6 in. The coefficient of consolidation (C_v) is 9.04×10^{-4} in.^{2}/min.

(a) How long will it take for 90% of the expected total settlement to take place?

(b) Compute the amount of settlement that will occur in 1 year.

(c) How long will it take for 1 in. of settlement to take place?

4-4 A 9-ft by 9-ft square footing to carry a total load of 300 tons is to be installed 6 ft below the ground surface on a sand stratum. Standard penetration tests were conducted on the site. The results of the tests were corrected for overburden pressures, and the corrected SPT values ($N_{\text{corrected}}$) are listed below:

Depth (ft)	Corrected SPT Values, $N_{\text{corrected}}$
2.5	25
5.0	28
7.5	27
10.0	30
12.5	28
15.0	23
17.5	24
20.0	28

No groundwater was encountered during subsurface exploration. Estimate the maximum expected settlement for the footing.

4-5 Assume the same conditions as in Problem 4-4 except that the groundwater table is located 8 ft below the ground level and the unit weight of the sand is 130 pcf. Estimate the maximum expected settlement for the footing.

4-6 A square footing 6 ft by 6 ft is to be installed 6 ft below the ground level on a sand stratum. Standard penetration tests were conducted on the construction site.

The results of the tests were corrected for overburden pressures, and the lowest average corrected N-value was determined to be 18. Assuming that groundwater was not encountered, determine the allowable soil pressure for a maximum of 1 in. of settlement.

4-7 Assume the same conditions as in Problem 4-6 except that the groundwater table is located 8 ft below the ground level and the unit weight of the sand is 118 pcf. Determine the allowable soil pressure for a maximum of 1 in. of settlement.

References

[1] WAYNE C. TENG, *Foundation Design*, Prentice-Hall, Inc., Englewood Cliffs, N.J., 1962.

[2] JOSEPH E. BOWLES, *Engineering Properties of Soils and Their Measurement*, 2nd ed., McGraw-Hill Book Company, New York, 1978.

[3] RALPH B. PECK, WALTER E. HANSEN, AND THOMAS H. THORNBURN, *Foundation Engineering*, 2nd ed., John Wiley & Sons, Inc., New York, 1974. Copyright © 1974, by John Wiley & Sons, Inc. Reprinted by permission of John Wiley & Sons, Inc.

[4] A. CASAGRANDE, "The Determination of the Pre-consolidation Load and Its Practical Significance," *Proc. First Int. Conf. Soil Mech., Cambridge, Mass.*, **3**, 60–64 (1936).

[5] J. H. SCHERTMANN, "The Undisturbed Consolidation Behavior of Clay," *Trans. ASCE*, **120**, 1201–1227 (1955).

[6] KARL TERZAGHI AND RALPH B. PECK, *Soil Mechanics in Engineering Practice*, John Wiley & Sons, Inc., New York, 1967. Copyright © 1967, by John Wiley & Sons, Inc. Reprinted by permission of John Wiley & Sons, Inc.

[7] J. H. SCHERTMANN, "Estimating the True Consolidation Behavior of Clay from Laboratory Test Results," *Proc. ASCE*, **79**, Separate 311 (1953). 26 pp.

[8] A. W. SKEMPTON, "Notes on the Compressibility of Clays," *Quart. J. Geol. Soc. Lond.*, **C**, 119–135, 1944.

[9] Abdel Rahman Sadik Said Bazaraa, "Use of the Standard Penetration Test for Estimating Settlements of Shallow Foundations on Sand," Ph.D. thesis, University of Illinois, 1967.

5

Shear Strength of Soil

5-1 INTRODUCTION

As a structural member, a piece of steel is capable of resisting compression, tension, and shear. Soil, however, like concrete and rock, is not capable of resisting high tension stresses (nor is it required to do so). It is capable of resisting compression to some extent; but in the case of excessive (failure producing) compression, failure usually occurs in the form of a shearing along some internal surface within the soil. Thus, the structural strength of soil is primarily a function of its shear strength, where shear strength of soil refers to the ability of the soil to resist sliding along internal surfaces within a mass of the soil.

Since the ability of soil to support an imposed load is determined by its shear strength, the shear strength of soil is of great importance in foundation design (Chap. 6), lateral earth pressure calculations (Chap. 9), slope stability analysis (Chap. 12), and in many other considerations. As a matter of fact, the shear strength of soil is of such great importance that it is a factor in most soil problems. The determination of shear strength is one of the most frequent, important problems in soil mechanics.

As explained in Sec. 1-7, soil gains its shear strength from two sources—internal friction and cohesion—as indicated by the Coulomb equation, Eq. (1-9), which is repeated here [1].

$$s = c + \sigma' \tan \phi \qquad (1\text{-}9)$$

where s = shear strength, psf

            c = cohesion, psf

$\sigma' =$ effective intergranular normal (perpendicular to the shear plane) pressure, psf

$\phi =$ angle of internal friction, deg

$\tan \phi =$ coefficient of friction

Cohesion (c) refers to strength gained from the ionic bond between grain particles and is predominant in cohesive soils. The angle of internal friction (ϕ) refers to strength gained from internal frictional resistance and is predominant in coarser-grained (cohesionless) soils. Cohesion (c) and the angle of internal friction (ϕ) might be referred to as the shear strength parameters. These two parameters can be evaluated for a given soil by standard laboratory and/or field tests (Sec. 5-2), thereby defining the relationship for shear strength (s) as a function of effective intergranular normal pressure (σ'). The latter term (σ') is not a soil parameter; it refers instead to the magnitude of the applied load.

As indicated in the preceding paragraph, the same two parameters affect the shear strength of both cohesive and cohesionless soils. However, the predominant parameter differs depending on whether a cohesive soil or a cohesionless soil is being considered. Accordingly, the study and analysis of shear strength of soil are normally done separately for cohesive soil and for cohesionless soil.

Field and laboratory methods for evaluating the shear strength parameters, from which shear strength can be determined, are presented in Sec. 5-2. The study and analysis of shear strength of cohesive soils are presented in Sec. 5-3, and those of cohesionless soils are presented in Sec. 5-4.

5-2 METHODS OF INVESTIGATING SHEAR STRENGTH

There are several methods of investigating the shear strength of soil. Discussed here are (1) unconfined compression test, (2) vane test, (3) direct shear test, and (4) triaxial compression test. The unconfined compression test can be used only to investigate cohesive soils, and the vane test can be used to investigate soft clays—particularly sensitive clays. The direct shear test and the triaxial test can be used to investigate both cohesive and cohesionless soils. As done previously in this book, only generalized discussions of the various test procedures are presented here; for specific instructions the reader is referred to a soil-testing manual.

Unconfined Compression Test

The unconfined compression test is perhaps the simplest, easiest, and least expensive test for investigating shear strength. It is quite similar to the usual determination of the compressive strength of concrete, where crushing a concrete cylinder is carried out solely by measured increases in end loading. A cylindrical cohesive soil sample is cut to a length of about $2\frac{1}{2}$ times the diameter of the sample. The sample is then placed in a compression testing machine (Fig. 5-1) and subjected to an axial load. The axial load is applied

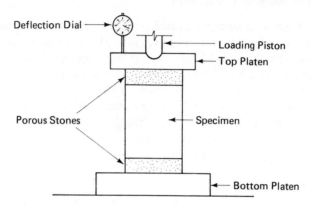

Deflection Dial

Loading Piston

Top Platen

Porous Stones

Specimen

Bottom Platen

FIGURE 5-1 Unconfined compression test apparatus. [2]

to produce axial strain at a rate of $\frac{1}{2}$ to 2%/min, and the resulting stress and strain are measured. The load is increased until the sample fails, and the highest compressive strength (called the unconfined compressive strength and denoted by q_u) is recorded. The cohesion [c in Eq. (1-9)] is taken to be $q_u/2$.

In the unconfined compression test, since there is no lateral support, the soil sample must be able to stand alone in the shape of a cylinder. A cohesionless soil (such as sand) cannot generally stand alone in this manner without lateral support; hence, this test procedure is usually limited to cohesive soils.

EXAMPLE 5-1

Given

1. A clay soil sample is subjected to an unconfined compression test.

2. The sample fails at a pressure of 2540 psf [i.e., unconfined compressive strength (q_u) = 2540 psf].

Required

Determine the cohesion of the clay soil.

Solution

$$\text{Cohesion} = \frac{\text{unconfined compressive strength}}{2}$$

or

$$c = \frac{q_u}{2}$$

$$\text{Cohesion } (c) = \frac{2540}{2} = 1270 \text{ psf}$$

Vane Test

The vane test, which was discussed in Sec. 2-5, can also be used to determine the shear strength of cohesive soils. This test can be used in the field to determine *in situ* shear strength for soft clay soil—particularly for sensitive clays (those that lose part of their strength when disturbed). The test can also be carried out in the laboratory on a cohesive soil sample.

Direct Shear Test

To carry out a direct shear test, a soil specimen is placed in a relatively flat box, which may be round or square (Fig. 5-2). A normal load of specific

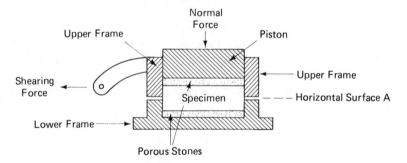

FIGURE 5-2 Typical direct shear box for single shear. [3]

(and constant) magnitude is applied. The box is "split" into two parts horizontally (see Fig. 5-2); and if one half is held while the other half is pushed with sufficient force, the soil will experience shear failure along horizontal surface *A*. This procedure is carried out in a direct shear apparatus (Fig. 5-3) and the particular normal load and the shear stress that produced shear failure are recorded. The soil specimen is then removed from the shear box and discarded, and another specimen of the same soil is placed in the shear box. A normal load different from (either higher or lower) the normal load used in the first test is applied to the second specimen, and a shear force is again applied with sufficient magnitude to cause shear failure. The normal load and the shear stress that produced shear failure are recorded for the second test.

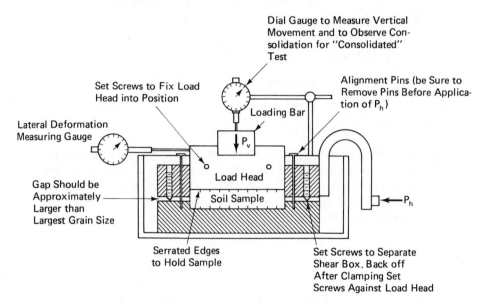

FIGURE 5-3 Direct shear apparatus. [2]

The results of these two tests must be plotted with normal stress (which is the total normal load divided by the cross-sectional area of the specimen) along the abscissa and shear stress (which is the shear stress producing shear failure) along the ordinate (see Fig. 5-4). (The same scale must be used along both the abscissa and the ordinate.) A straight line is drawn connecting these

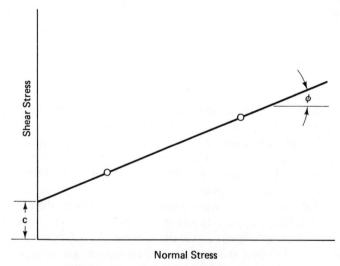

FIGURE 5-4 Shear diagram for direct shear test.

two plotted points, with the straight line extended to intersect the ordinate (i.e., the vertical axis). The angle between this straight line and a horizontal line (ϕ in Fig. 5-4) gives the value of the angle of internal friction [ϕ in Eq. (1-9)], and the value of shear stress where the straight line intersects the ordinate (c in Fig. 5-4) gives the value of cohesion [c in Eq. (1-9)]. These values of ϕ and c can be used in Eq. (1-9) to determine the shear strength of the particular soil for any load (i.e., for any normal pressure, σ').

In theory, it is adequate to have only two points to define the straight-line relationship of Fig. 5-4. In practice however, it is better to have three (or more) such points through which the best straight line can be drawn. This means, of course, that three (or more) separate tests must be made on three (or more) specimens from the same soil sample.

EXAMPLE 5-2

Given

1. A series of direct shear tests was performed on a soil. Each test was carried out until the soil sample sheared.

2. Laboratory data for the tests are listed below.

Specimen Number	Normal Stress (ksf)	Shearing Stress (ksf)
1	0.25	0.35
2	0.50	0.56
3	1.0	0.94

Required

Determine the cohesion and the angle of internal friction of the soil.

Solution

Plot the data given on a shear diagram (see Fig. 5-5). The scales on both ordinate and abscissa of the shear diagram should be equal. Connect the plotted points by a straight line and note that the line makes an angle of 38° with the horizontal and intersects the vertical axis (ordinate) at 0.16 ksf. Therefore, cohesion (c) = 0.16 ksf and angle of internal friction (ϕ) = 38°.

The direct shear test is a relatively simple test to determine the shear strength parameters of soils. However, in this test shear failure is forced to occur along or across a predetermined plane (surface A in Fig. 5-2), which is not necessarily the weakest plane of the soil specimen tested. Since the development of the much better triaxial test (discussed next), the use of the direct shear test has decreased.

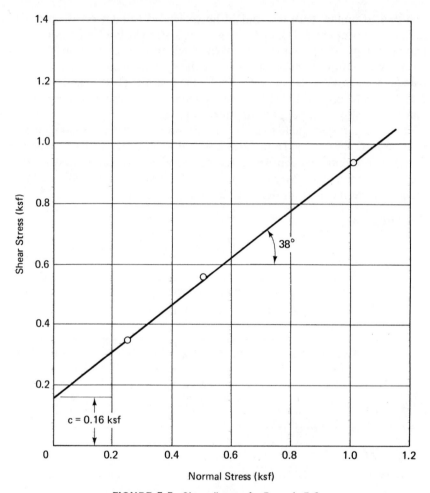

FIGURE 5-5   Shear diagram for Example 5-2.

Triaxial Compression Test

The triaxial compression test is carried out in a manner somewhat similar to the unconfined compression test in that a cylindrical soil sample is subjected to a vertical (axial) load. The major difference is that, unlike the unconfined compression test where there is no confining (lateral) pressure, the triaxial test is carried out with confining (lateral) pressure present. The lateral pressure is made possible by enclosing the sample in a chamber (see Fig. 5-6) and introducing water or compressed air into the chamber, with the water or compressed air surrounding the soil sample.

To carry out a test, a cylindrical soil specimen having a length about $2\frac{1}{2}$ times its diameter is wrapped in a rubber membrane and placed in the triaxial

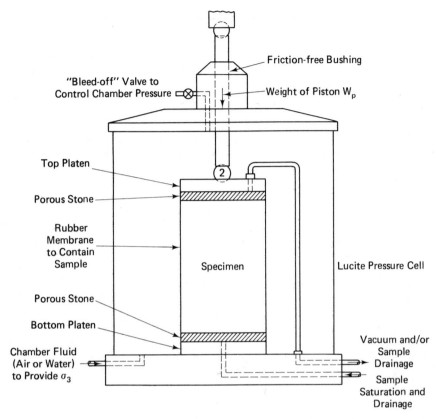

FIGURE 5-6 Schematic diagram of the triaxial chamber. [2]

chamber, and a specific (and constant) lateral pressure is applied by means of water or compressed air within the chamber. A vertical (axial) load is then applied externally and steadily increased until the specimen fails. The externally applied axial load that causes the specimen to fail and the lateral pressure are recorded. As in the direct shear test, it is necessary to remove the soil specimen and discard it, and then to place another specimen of the same soil sample in the triaxial chamber. The procedure described above is repeated for the new specimen for a different (either higher or lower) lateral pressure. The axial load at failure and the lateral pressure are recorded for the second test.

The lateral pressure is designated as σ_3. However, it is applied not only to the sides of the cylindrical specimen but also to the ends. It is therefore called the *minor principal stress*. The externally applied axial load at failure divided by the cross sectional area of the test specimen is designated as Δp, and it is called the *deviator stress*. The total vertical (axial) pressure causing failure is the sum of the minor principal stress (σ_3) and the deviator stress (Δp). This total vertical (axial) pressure is designated as σ_1, and it is called the *major principal stress*. In equation form,

$$\sigma_1 = \sigma_3 + \Delta p \qquad (5\text{-}1)$$

The results of the triaxial compression tests can be plotted in the following manner. Using the results of one of the triaxial tests, a point is located along the abscissa at a distance of σ_3 from the origin. This point is denoted by A in Fig. 5-7, and it is indicated as being located along the abscissa at a distance of $(\sigma_3)_1$ from the origin. It is also necessary to locate another point along the

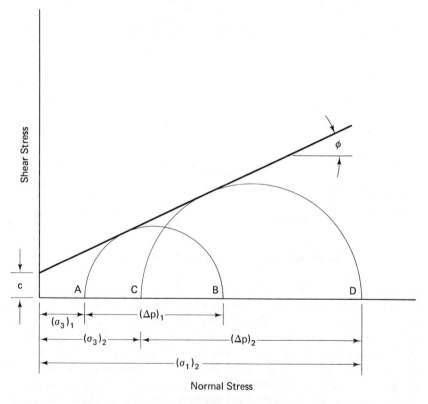

FIGURE 5-7 Shear diagram for triaxial compression test.

abscissa at a distance of σ_1 from the origin. This point can be located either by measuring the distance σ_1 from the origin or by measuring the distance Δp from point A (the point located at a distance σ_3 from the origin). This point is denoted by B in Fig. 5-7, and it is indicated as being located along the abscissa at a distance $(\Delta p)_1$ from point A. Using AB as a diameter, a semicircle is constructed as shown in Fig. 5-7. (This is known as "Mohr circle.") This procedure is repeated using the data obtained from the triaxial test of the other specimen of the same soil sample. Thus point C is located along the abscissa at a distance of $(\sigma_3)_2$ from the origin, and point D is located along the abscissa at a distance $(\Delta p)_2$ from point C. Using CD as a diameter, another semicircle is constructed. The next step is to draw a straight line that is tangent to the semicircles, as shown in Fig. 5-7. As in the direct shear test (Fig. 5-4), the angle between this straight line and a horizontal line (ϕ in Fig. 5-7) gives the value of the angle of internal friction [ϕ in Eq. (1-9)], and the value of stress where the straight line intersects the ordinate (c in Fig. 5-7) gives the value of cohesion [c in Eq. (1-9)]. The same scale must be used along both the abscissa and the ordinate.

As in the direct shear test, it is adequate, in theory, to have only two Mohr circles to define the straight-line relationship of Fig. 5-7. In practice, however, it is better to have three (or more) such semicircles that can be used to draw the best straight line. This means, of course, that three (or more) separate tests must be performed on three (or more) specimens from the soil sample.

EXAMPLE 5-3

Given

1. Triaxial compression tests of three specimens of the same soil sample were performed in a soil laboratory. Each test was carried out until the sample failed.

2. The data obtained in the tests are tabulated below.

Specimen Number	Minor Principal Stress, σ_3 (Confining Pressure) (ksf)	Major Principal Stress, σ_1 $(\sigma_3 + \Delta p)$ (ksf)
1	2	11.0
2	4	15.2
3	6	18.8

Required

Determine the cohesion and the angle of internal friction of the soil.

Solution

As shown in Fig. 5-8, draw three Mohr circles. Each circle starts at a minor principal stress (σ_3) and has a diameter equal to the difference between major principal stress (σ_1) and minor principal stress (σ_3) for one test [maximum deviator stress (Δp) = $\sigma_1 - \sigma_3$]. Then draw the Mohr envelope of ruptures as nearly as possible tangent to all three circles. The cohesion of the soil,

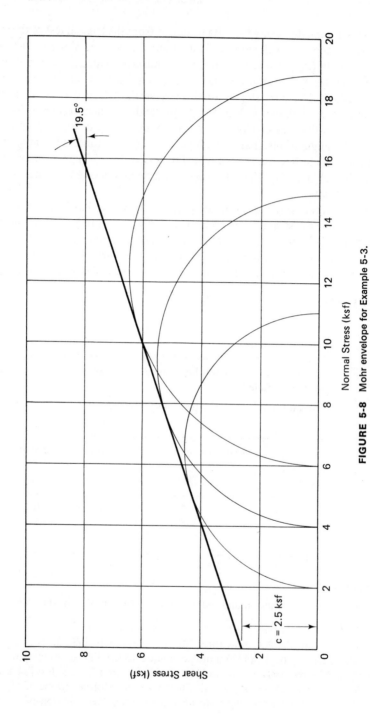

FIGURE 5-8 Mohr envelope for Example 5-3.

118

which is indicated by the intersection of the Mohr envelope and the vertical axis (ordinate) is 2.5 ksf. The angle of internal friction of the soil, which is the angle that the envelope makes with the horizontal axis, is 19.5°.

EXAMPLE 5-4

Given

1. A sample of dry cohesionless soil is subjected to a triaxial test.
2. The angle of internal friction is estimated to be 37°.

Required

If the minor principal stress (σ_3) is 14 psi, at what values of the maximum deviator stress (Δp) and major principal stress (σ_1) is the sample likely to fail?

Solution

All samples of dry cohesionless soils will have cohesions equal to zero. Therefore, the Mohr envelope must go through the origin. Draw a Mohr envelope starting at the origin for $\phi = 37°$. Then draw the Mohr circle, starting at a minor principal stress (σ_3) of 14 psi and tangential to the Mohr envelope (see Fig. 5-9). It can now be determined that the maximum deviator

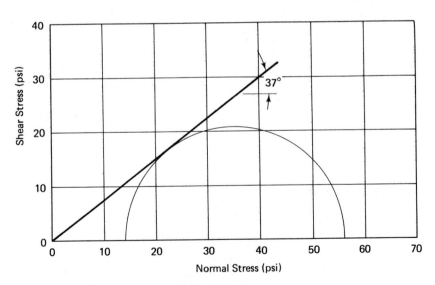

FIGURE 5-9 Mohr circle for Example 5-4.

stress (Δp) = 42.3 psi (note that the maximum deviator stress equals the diameter of the Mohr circle), and the major principal stress = $\Delta p + \sigma_3$ = 42.3 + 14 = 56.3 psi.

As will be shown in Chap. 6, the angle of internal friction (ϕ) can be approximated for a cohesionless soil, utilizing the results of a standard penetration test (SPT).

5-3 SHEAR STRENGTH OF COHESIVE SOILS

When load is applied to a saturated, or nearly saturated, cohesive soil (most clays in their natural condition are close to full saturation), the water in the voids of the cohesive soil carries the load first and consequently prevents the relatively small cohesive soil particles from coming into contact to develop frictional resistance. At that time, the shear strength of the cohesive soil consists only of cohesion (i.e., $s = c$). As time goes on, the water in the voids of cohesive soils is slowly expelled from the voids, and the soil particles come together and offer frictional resistance. This increases the shear strength from $s = c$ to $s = c + \sigma' \tan \phi$ [see Eq. (1-9)]. Because the permeability of cohesive soil is very low, the process of water being expelled or extruded from the voids is very slow, perhaps occurring over a period of years. What all of this means is that, immediately after a structure is built (i.e., immediately upon load application), the shear strength of a saturated cohesive soil consists only of cohesion. Therefore, in foundation design problems, the bearing capacity of cohesive soil should be estimated based on the assumption that soil behaves as if the angle of internal friction (ϕ) is equal to zero, and shear strength is equal to cohesion. Such a design practice should be adequate at construction time, and subsequent increase in shear strength should give an added factor of safety to the foundation [1].

For most cohesive soils, the cohesion is estimated from the results of unconfined compression tests. However, for soft and/or sensitive clay, the cohesion is commonly obtained from the results of field or laboratory vane tests (Sec. 5-2).

5-4 SHEAR STRENGTH
OF COHESIONLESS SOILS

Because of relatively large particle size, cohesionless soils possess virtually no cohesion. This is because the large particles have no tendency to stick together. The large particles do, however, develop significant frictional resistance, including sliding and rolling friction as well as interlocking of the grains. This will give significant values of the angle of internal friction (ϕ). With no cohesion ($c = 0$), Eq. (1-9) reverts to $s = \sigma' \tan \phi$.

As related in the preceding section, the extrusion of water from void space is an extremely slow process for a cohesive soil. Accordingly, the most critical condition with regard to shear strength usually occurs at construction time or upon application of load. In the case of cohesionless soils, any water contained in void space at construction time or upon application of load will be driven out almost immediately, because of the high permeability of the soil. Thus, the shear strength of the cohesionless soil remains more or less the same throughout the life of the structure.

The angle of internal friction (ϕ) of cohesionless soils can be obtained from laboratory or field tests (Sec. 5-2). However, ϕ can also be estimated based on the correlation between corrected SPT (standard penetration test) values and ϕ given by Peck et al. [4]. This correlation is shown in Fig. 2-9. To use this graph, one enters at the upper right with the corrected SPT value, moves horizontally to the curve marked N, then moves vertically down to the abscissa where the value of ϕ is read.

5-5 PROBLEMS

5-1 A series of direct shear tests was performed on a soil. Each test was carried out until the soil sample sheared. Laboratory data for the tests are tabulated below. Determine the cohesion and the angle of internal friction of the soil.

Specimen Number	Normal Stress (lb/ft^2)	Shearing Stress(lb/ft^2)
1	200	450
2	400	520
3	600	590
4	1000	740

5-2 The data shown below were obtained in triaxial compression tests of three identical soil samples. Find the cohesion and the angle of internal friction of the soil.

Specimen Number	Minor Principal Stress, σ_3 ($lb/in.^2$)	Major Principal Stress, σ_1 ($lb/in.^2$)
1	5	23.0
2	10	38.5
3	15	53.6

5-3 A cohesionless soil sample was subjected to a triaxial test. The sample failed when the minor principal stress (all around or confining pressure) was 1200 lb/ft^2 and the maximum deviator stress (Δp) was 3000 lb/ft^2. Find the angle of internal friction for this soil.

5-4 A sample of dry cohesionless soil is known to have an angle of internal friction of 35°. If the minor principal stress (σ_3) is 15 psi, at what values of the maximum deviator stress (Δp) and major principal stress (σ_1) is the sample likely to fail?

5-5 A cohesive soil sample is subjected to an unconfined compression test in a soil laboratory. The sample fails at a pressure of 3850 psf [i.e., unconfined compressive strength (q_u) = 3850 psf]. Determine the cohesion of the soil.

References

[1] WAYNE C. TENG, *Foundation Design*, Prentice-Hall, Inc., Englewood Cliffs, N.J., 1962.

[2] JOSEPH E. BOWLES, *Engineering Properties of Soils and Their Measurement*, 2nd ed., McGraw-Hill Book Company, New York, 1978.

[3] *Standard Specifications for Transportation Materials and Methods of Sampling and Testing, Part I, Specifications*, 12th ed., AASHTO, 1978.

[4] RALPH B. PECK, WALTER E. HANSEN, AND THOMAS H. THORNBURN, *Foundation Engineering*, 2nd ed., John Wiley & Sons, Inc., New York, 1974. Copyright © 1974, by John Wiley & Sons, Inc. Reprinted by permission of John Wiley & Sons, Inc.

6

Shallow Foundations

6-1 INTRODUCTION

The word "foundation" might be defined in general as "that which supports something." Many universities, for example, have an "athletic foundation," which supports in part the school's sports program. In the context of this book, foundation normally refers to that which supports a structure, such as a column or wall, along with the loads carried by the structure.

Foundations may be characterized as being either "shallow" or "deep." Shallow foundations are those located just below the lowest part of the superstructures which they support; deep foundations extend considerably deeper into the earth. In the case of shallow foundations, the means of support is usually either a "footing," which is often simply an enlargement of the base of the column or wall it supports, or a "mat" or "raft foundation," in which a number of columns are supported by a single slab. The remainder of this chapter deals with footings. In the case of deep foundations, the means of support is usually either a pier, a caisson, or a group of piles. These will be covered in Chaps. 7 and 8.

An individual footing is shown in Fig. 6-1a. For purposes of analysis, a footing such as this may be thought of as a simple flat plate or slab, usually square in plan, acted on by a concentrated load (the column) and a distributed load (soil pressure) (see Fig. 6-1b). The enlargement of the size of the footing (compared to the column that it supports) gives an increased contact area between footing and soil, and the increased area serves to reduce the pressure on the soil to an allowable amount, thereby preventing excessive settlement or other bearing failure of the foundation.

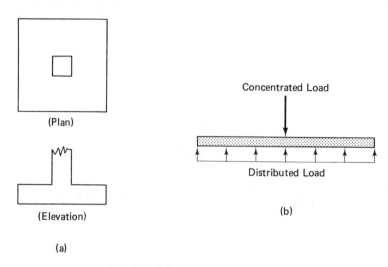

FIGURE 6-1 An individual footing.

Footings may be classified in several ways. For example, the footing depicted in Fig. 6-1a is an *individual footing*. Sometimes one larger footing may support two or more columns, as shown in Fig. 6-2a. This is known as a *combined footing*. If a footing is extended in one direction to support a long structure such as a wall, it is called a *continuous footing*, or a *wall footing* (Fig. 6-2b). Two or more footings joined by a beam (called a strap) is called a *strap footing* (Fig. 6-2c). A large slab supporting a number of columns not all of which are in a straight line is called a *mat* or *raft foundation* (Fig. 6-2d).

Foundations must be designed to satisfy three general criteria:

1. They must be located properly (both vertical and horizontal orientation) so as not to be adversely affected by outside influences.
2. They must be safe from bearing capacity failure (collapse).
3. They must be safe from excessive settlement.

Specific procedures for designing footings are given in the remainder of this chapter. For initial orientation and for future quick reference, the following steps for designing footings are offered at this point:

1. Calculate loads acting on the footing—Sec. 6-2.
2. Obtain a soil profile or soil profiles along with pertinent field and laboratory measurements and testing results—Chap. 2.
3. Determine the depth and location of footings—Sec. 6-3.
4. Determine the bearing capacity of the supporting soil—Sec. 6-4.

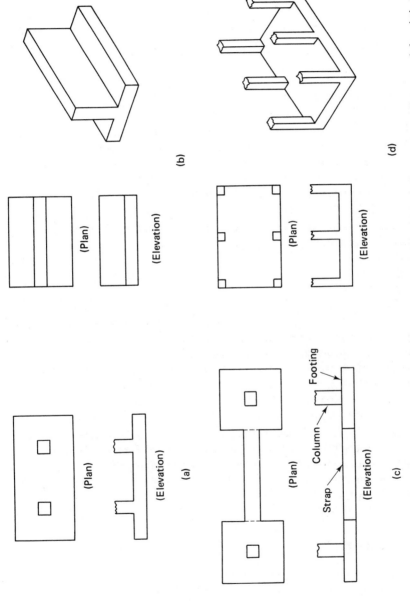

FIGURE 6-2 Classification of footings: (a) combined footing; (b) wall footing; (c) strap footing; (d) mat or raft foundation.

(Plan)

(Elevation)

(a)

(Plan)

(Elevation)

(b)

Footing

Column

Strap

(Plan)

(Elevation)

(c)

(Plan)

(Elevation)

(d)

5. Determine the size of footing—Sec. 6-5.

6. Compute the footing contact pressure and check stability against sliding and overturning—Sec. 6-6.

7. Estimate the total and differential settlements—Chap. 4 and Sec. 6-7.

8. Design the footing structure—Sec. 6-8.

6-2 LOADS ON FOUNDATIONS [1]

When designing any structure, whether it is a steel beam or column, a floor slab, a foundation, or whatever, it is of basic and utmost importance that an accurate estimation (computation) of all loads acting on the structure be made. In general, a structure may be subjected upon construction or sometime in the future to some or all of the following loads, forces, and pressures: (1) dead load, (2) live load, (3) wind load, (4) snow load, (5) earth pressure, (6) water pressure, and (7) earthquake forces. These will be discussed in the following paragraphs.

Dead Load

Dead load refers to the overall weight of the structure itself. It includes the weight of materials permanently attached to the structure (such as flooring) and fixed service equipment (such as air-conditioning equipment). Dead load can be calculated if sizes and types of structural material are known. This presents a problem, however, because the weight of a structure is not known until its size is known, and its size cannot be known until it has been designed based (in part) on its weight. A normal procedure is to estimate the dead load initially, use this estimated load (along with other types of loads, such as live load, wind load, etc.) to size the structure, and then compare the weight of the specified structure with the estimated weight. If the computed weight differs appreciably from the estimated weight, the design procedure should be repeated, using a revised estimated weight. In the case of a footing design, both the weights of the footing and the superstructure it supports can be estimated by this procedure.

Live Load

Live load refers to weights of applied bodies that are not a permanent part of the structure. These may be applied to the structure during part of its useful life (such as people, warehouse goods) or during all of its useful life (such as furniture). Because of the nature of live loads, it would be virtually impossible in most cases to calculate live load directly. Instead, live loads to be used in structural design are usually specified by local building codes. For example, a state building code may specify a minimum live loading of 100 psf for restaurants and 80 psf for floor in office buildings.

Wind Loads

Wind loads, which are not considered as live loads, may act on all exposed surfaces of structures. Additionally, overhanging parts of buildings may be subject to uplift pressure as a result of wind. Like live loads, design wind loads are usually calculated based on local building codes. For example, a state building code may specify a design wind loading for a particular county of 15 psf for buildings less than 30 ft tall and 40 psf for buildings taller than 1200 ft, with a sliding scale in between.

Snow Loads

Snow loads result from accumulation of snow on roofs and exterior flat surfaces. The unit weight of snow varies, but it weighs about 6 pcf on the average. Thus, it is obvious that an accumulation of several feet of snow over a large roof area results in a very heavy load. (Two feet of snow over a 50-ft by 50-ft roof would be about 15 tons.) Design snow loads are also usually based on local building codes. A state building code may specify a minimum snow loading of 30 psf for a specific county.

Earth Pressure

Earth pressure produces a lateral force that acts against the portion of substructure lying below ground or fill level (see Fig. 6-3a). It is normally treated as dead load.

Water Pressure

Water pressure may produce a lateral force similar in nature to that produced by earth pressure. Water pressure may also produce a force that acts upward (hydrostatic uplift) on the bottom of the structure. These forces are illustrated in Fig. 6-3b. Lateral water pressure is generally balanced, but hydrostatic uplift is not. It must be counteracted by the dead load of the structure, or else some provision must be made to anchor the structure.

Earthquakes

Earthquakes produce forces that may act laterally, vertically, or torsionally on a structure in any direction. A building code should be consulted for the specification of earthquake forces.

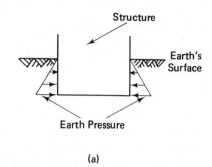

Structure

Earth's Surface

Earth Pressure

(a)

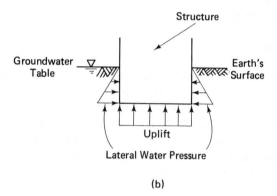

Structure

Groundwater Table

Earth's Surface

Uplift

Lateral Water Pressure

(b)

FIGURE 6-3 (a) Earth pressure; (b) water pressure.

6-3 DEPTH AND LOCATION OF FOUNDATIONS [2]

As related previously (Sec. 6-1), foundations must be located properly (both vertical and horizontal orientation) so as not to be adversely affected by outside influences. Outside influences would include adjacent structures; water, including frost and groundwater; significant soil volume change; and underground defects (caves, for example). Thus, depth and location of foundations are dependent on the following factors:

1. Frost action.
2. Significant soil volume change.
3. Adjacent structures and property lines.
4. Groundwater.
5. Underground defects.

These factors will be discussed in the following paragraphs.

Frost Action

In areas where the air temperature falls below the freezing point, the moisture in the soil near the surface of the ground may freeze. When the temperature subsequently rises above the freezing point, any frozen moisture may melt. As the soil moisture freezes and melts, it alternately expands and contracts. Repeated expansion and contraction of soil moisture beneath a footing may cause it to be lifted during cold weather and dropped during warmer weather. Such a sequence usually cannot be tolerated by the structure.

The general solution to prevention of frost action on footings is to place the foundation below the depth of soil that is expected ever to be penetrated by frost. This depth of frost penetration varies from 4 ft or more in some northern states (Maine, Minnesota) to zero in parts of some southern states (Florida, Texas). Since frost penetration varies with location, local building codes often dictate minimum depths of footings.

Significant Soil Volume Change

Many soils, particularly certain clays having high plasticity, shrink significantly upon drying and swell significantly upon wetting. This volume change is greatest near the ground surface and decreases with increasing depth. The specific depth and volume change relationship for a particular soil is dependent on the type of soil and the level of groundwater. The volume change is usually insignificant below a depth of from 5 to 10 ft and does not occur below the groundwater table. Generally speaking, the soil beneath the center of a structure is more protected from sun and precipitation, and therefore, moisture change and resulting soil movement are smallest there. On the other hand, the soil beneath the edges of a structure is less protected, and moisture change and consequent soil movement are greatest there.

As in the case of frost action, significant soil volume change beneath a footing may cause alternate lifting and dropping of the footing. Possible means of avoidance include placing the footing below all strata that are subject to significant volume changes (those soils with plasticity indices over 30), placing it below the zone of volume change, and placing it below any objects that could affect moisture content unduly (such as roots, steam lines, etc.).

Adjacent Structures and Property Lines

Adjacent structures and property lines often affect the horizontal location of a footing. Adjacent (existing) structures may be damaged by construction of new foundations nearby as a result, for example, of vibration, shock resulting from blasting, undermining by nearby excavation, or lowering the water table. After new foundations have been constructed, the (new) load they place on the soil may cause settlement of previously existing structures as a result of new stress patterns in the surrounding soil.

Since damage to existing structures by new construction may result in problems of liability, new structures should be located and designed very carefully. In general, the deeper the new foundation and the closer to the old structure, the greater the potential for damage to the old structure. Accordingly, old and new foundations should be separated as much as is practical. This is particularly true if the new foundation will be lower than the old one. A general rule is that a straight line drawn downward and outward at a 45° angle from the end of the bottom of any new (or existing) higher footing should not intersect any existing (or new) lower footing (see Fig. 6-4).

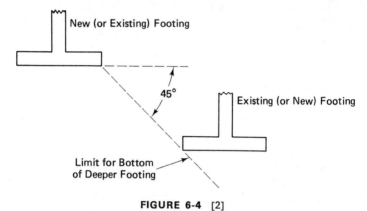

FIGURE 6-4 [2]

Special care must be exercised in placing a footing at or near a property line. One reason is that, since a footing is wider than the structure it supports, it is possible for part of the footing to extend across a property line and encroach on adjacent land, although the structure supported by the footing does not do so (see Fig. 6-5). Also, excavation for a footing at or near a prop-

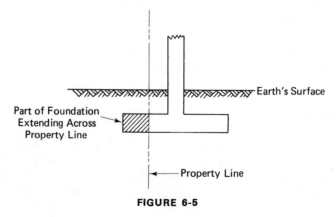

FIGURE 6-5

erty line may have a harmful effect (cave-in, for example) on adjacent land. Either of these cases could result in liability problems; hence, much care should be exercised when a footing is required near a property line.

Groundwater

The presence of groundwater within the soil immediately around a footing is undesirable for several reasons. First, footing construction below the groundwater level is difficult and expensive. Generally, the area must be drained prior to construction. Second, groundwater around a footing can reduce the strength of soils by reducing their ability to carry foundation pressures. Third, groundwater around a footing may cause hydrostatic uplift problems; fourth, frost action may increase; and fifth, if the groundwater reaches the lowest floor of the structure, waterproofing problems are encountered. For these reasons, footings should be placed above the groundwater level whenever practical to do so.

Underground Defects

Footing location is also affected by the presence of underground defects. These include faults, caves, and mines, as well as man-made discontinuities such as sewer lines and underground cables and utilities. Minor breaks in bedrock seldom are a problem unless they are active. Structures should never be built on or near tectonic faults that may slip. Certainly, foundations placed directly above a cave or mine should be avoided if at all possible. The man-made discontinuities listed above are often encountered, and obviously foundations should not be placed above them. When they are encountered where a footing is desired, either they or the footing should be relocated. As a matter of fact, a survey of underground utility lines should be made prior to excavation for a foundation in order to avoid damage (or even an explosion) to the utility lines during excavation.

6-4 BEARING CAPACITY ANALYSIS

The conventional method of designing foundations is based on the concept of "bearing capacity." One meaning of the verb "to bear" is to support or to hold up. Generally, therefore, bearing capacity refers to the ability of a soil to support or to hold up a foundation and structure. The "ultimate bearing capacity" of a soil refers to the loading per unit area that will just cause a shear failure in the supporting soil. It is given the symbol q_{ult}. The "allowable bearing capacity" (symbol q_a) refers to the load per unit area that the soil is able to support without unsafe movement. It is the "design" bearing capacity. The allowable loading is equal to the allowable bearing capacity multiplied by the area of contact between foundation and soil. The allowable bearing capacity is equal to the ultimate bearing capacity divided by the factor of safety. A factor of safety of from 2.5 to 3 is commonly applied to the value of q_{ult}. Care must be taken to ensure that a footing design is safe with regard to (1) foundation failure (collapse) and (2) excessive settlement.

The following equations for calculating ultimate bearing capacity were developed by Terzaghi [3]:

Continuous footings (width B):

$$q_{ult} = cN_c + \gamma D_f N_q + 0.5\gamma B N_\gamma \tag{6-1}$$

Circular footings (radius R):

$$q_{ult} = 1.2cN_c + \gamma D_f N_q + 0.6\gamma R N_\gamma \tag{6-2}$$

Square footings (width B):

$$q_{ult} = 1.2cN_c + \gamma D_f N_q + 0.4\gamma B N_\gamma \tag{6-3}$$

The terms in these equations are

q_{ult} = ultimate bearing capacity, psf

c = cohesion of soil, psf

N_c, N_q, N_γ = Terzaghi's bearing capacity factors

γ = effective unit weight of soil, pcf

D_f = depth of footing, or distance from ground surface to base of footing, ft

B = width of square footing, ft

R = radius of a circular footing, ft

Equations (6-1) through (6-3) are applicable for both cohesive and cohesionless soils. Values of Terzaghi's bearing capacity factors (N_c, N_q, N_γ) may be obtained by using the curves given in Fig. 6-6. Dense sand and stiff clay produce what is called "general shear," and the solid lines of Fig. 6-6 are used along with the angle of internal friction (ϕ) to determine values of N_c, N_q, and N_γ. Loose sand and soft clay produce what is called "local shear," and the dashed lines are used to determine values of N_c', N_q', and N_γ'. In the latter case (loose sand and soft clay), the term c (cohesion) in Eqs. (6-1) through (6-3) is replaced by c', which is equal to $\frac{2}{3}c$. Thus, the terms c', N_c', N_q', and N_γ' are used in Eqs. (6-1) through (6-3) in place of the respective unprimed terms for loose sand and soft clay.

In the case of cohesive soils, the shear strength of the soil is most critical

<u>Loaded Strip, Width B</u>

  Load per unit area of footing

  General shear failure: $q_d = cN_c + \gamma D_f N_q + \frac{1}{2}\gamma B N_\gamma$

  Local shear failure   : $q_d' = \frac{2}{3}cN_c' + \gamma D_f N_q' + \frac{1}{2}\gamma B N_\gamma'$

<u>Square Footing, Width B</u>

  Load per unit area : $q_{ds} = 1.2cN_c + \gamma D_f N_q + 0.4\gamma B N_\gamma$

Unit weight of earth $= \gamma$

Unit shear resistance,

  $s = c + \sigma \tan \phi$

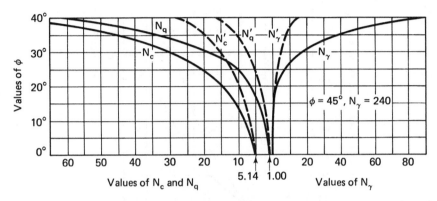

FIGURE 6-6 Chart showing relationship between ϕ and bearing capacity factors (values of N_γ after Meyerhof 1955). [3, 4]

just after construction or as load is first applied, at which time the shear strength is assumed to consist only of cohesion. In this case, ϕ (angle of internal friction) is assumed equal to zero [1]. There are several means of determining the cohesion [c term in Eqs. (6-1) through (6-3)]. One is to use the unconfined compression test for ordinary sensitive or insensitive clay. In this case, c is equal to one-half of the unconfined compressive strength (i.e., $\frac{1}{2}q_u$) (see Chap. 5). For sensitive clay, a field vane test may be used to evaluate cohesion (see Chap. 2).

    In the case of cohesionless soils, the c term in Eqs. (6-1) through (6-3) is zero. The value of ϕ (needed in order to use Fig. 6-6 to determine N_c, N_q, and N_γ) may be determined by several methods. One is to use the corrected standard penetration test values (see Chap. 2) and the curves shown in Fig. 6-7. One enters the graph at the upper right with the corrected SPT value, moves horizontally to the curve marked N, then moves vertically down to the abscissa to read the value of ϕ. The value of ϕ can be used with the curves in Fig. 6-6 to determine values of N_q and N_γ. Or, values of N_q and N_γ may be determined using Fig. 6-7 by projecting vertically down from the curve marked N to the curves marked N_q and N_γ, then projecting horizontally over to the ordinate to read values of N_q and N_γ, respectively. It is not necessary to determine a value of N_c, since c is equal to zero in the case of cohesionless soil, and thus the cN_c terms of Eqs. (6-1) through (6-3) are zero.

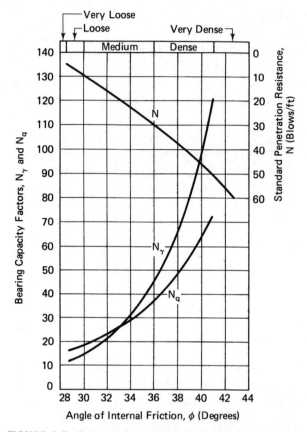

FIGURE 6-7 Curves showing the relationship between bearing capacity factors and ϕ, as determined by theory, and rough empirical relationship between bearing capacity factors or ϕ and values of standard penetration resistance N. [5]

The three example problems that follow demonstrate the application of the Terzaghi bearing capacity formulas [that is, Eqs. (6-1) through (6-3)]. Example 6-1 deals with a wall footing in stiff clay. Example 6-2 deals with a square footing in a stiff cohesive soil. Example 6-3 deals with a square footing in a dense cohesionless soil.

EXAMPLE 6-1

Given

1. A strip of wall footing 3.5 ft wide is supported in a uniform deposit of stiff clay.

2. Unconfined compressive strength of this cohesive soil (q_u) = 2.8 ksf.

3. Unit weight of the soil $(\gamma) = 130$ pcf.

4. Groundwater was not encountered during the subsurface soil exploration.

5. The depth of wall footing $(D_f) = 2$ ft (see Fig. 6-8).

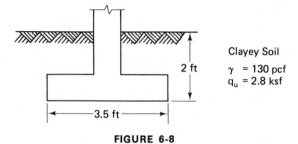

Clayey Soil
$\gamma = 130$ pcf
$q_u = 2.8$ ksf

2 ft

3.5 ft

FIGURE 6-8

Required

1. The ultimate bearing capacity of this foundation.

2. The allowable wall load in kips/ft of length of wall, using a factor of safety of 3.

Solution

1. For continuous wall footing,

$$q_{ult} = cN_c + \gamma D_f N_q + 0.5\gamma B N_\gamma \qquad (6\text{-}1)$$

$$c = \frac{q_u}{2} = \frac{2.8}{2} = 1.4 \text{ ksf}$$

$$\gamma = 0.13 \text{ kips per cubic foot (kcf)}$$

$$D_f = 2 \text{ ft}$$

Using $c > 0$, $\phi = 0$ analysis for cohesive soil; when $\phi = 0$, Fig. 6-6 gives

$$N_c = 5.14$$
$$N_q = 1.0$$
$$N_\gamma = 0$$

$$q_{ult} = (1.4)(5.14) + (0.13)(2)(1) + (0.5)(0.13)(3.5)(0) = 7.46 \text{ ksf}$$

2. $q_a = \dfrac{7.46}{3} = 2.49 \text{ ksf}$

Allowable wall loading $= q_a \times B \times$ unit ft length of wall
$= (2.49)(3.5)(1) = 8.7$ kips/ft of length of wall

EXAMPLE 6-2

Given

1. A square footing with 5-ft sides is located 4 ft below the ground surface.

2. The groundwater table is at very great depth and its effect may be ignored.

3. The subsoil consists of a thick deposit of stiff cohesive soil, with unconfined compressive strength (q_u) equal to 3000 psf.

4. Unit weight (γ) of this cohesive soil is 120 pcf (see Fig. 6-9).

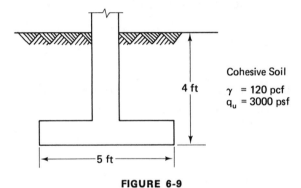

4 ft

Cohesive Soil
γ = 120 pcf
q_u = 3000 psf

5 ft

FIGURE 6-9

Required

The allowable bearing capacity using a factor of safety of 3.0.

Solution

Since the supporting stratum is stiff clay, general shear condition is evident in this case. For a square footing,

$$q_{ult} = 1.2cN_c + \gamma D_f N_q + 0.4\gamma BN_\gamma \tag{6-3}$$

$$c = \frac{q_u}{2} = \frac{3000}{2} = 1500 \text{ psf}$$

$$\gamma = 120 \text{ pcf}$$

$$D_f = 4 \text{ ft}$$

Using $c > 0$ and $\phi = 0$ analysis for cohesive soil; when $\phi = 0$, Fig. 6-6 gives

$$N_c = 5.14$$

$$N_q = 1.0$$

$$N_\gamma = 0$$

$B = 5\,\text{ft}$ (given)

$q_{\text{ult}} = (1.2)(1500)(5.14) + (120)(4)(1) + (0.4)(120)(5)(0) = 9730\ \text{psf}$

$q_a = \dfrac{q_{\text{ult}}}{\text{F.S.}} = \dfrac{9730}{3} = 3240\ \text{psf}$

EXAMPLE 6-3

Given

1. A column footing 6 ft by 6 ft is buried 5 ft below the ground surface in a dense cohesionless soil.
2. The results of laboratory and field tests of the soil are as follows:
   a. Unit weight of soil $(\gamma) = 128$ pcf.
   b. The average corrected standard penetration test value beneath the footing $= 30$.
   c. Groundwater was not encountered during the subsurface soil exploration.
3. The footing is to carry a total load of 300 kips, including column load, weight of footing, and weight of soil surcharge (see Fig. 6-10).

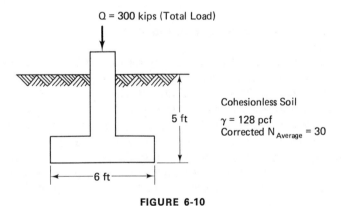

Q = 300 kips (Total Load)

Cohesionless Soil
γ = 128 pcf
Corrected N_{Average} = 30

5 ft

6 ft

FIGURE 6-10

Required

The factor of safety against bearing capacity failure (collapse).

Solution

Since the supporting stratum is dense cohesionless soil, general shear condition is evident. Hence, the Terzaghi bearing capacity formula for a square footing is used with $c = 0$, $\phi > 0$.

For a square footing

$$q_{\text{ult}} = 1.2cN_c + \gamma D_f N_q + 0.4\gamma B N_\gamma \tag{6-3}$$

$$c = 0 \quad \text{(cohesionless soil)}$$

$$\gamma = 128 \text{ pcf}$$

$$D_f = 5 \text{ ft}$$

$$B = 6 \text{ ft}$$

From Fig. 6-7, with corrected $N_{\text{average}} = 30$, $\phi = 36°$. Then enter with $\phi = 36°$ into Fig. 6-6, and the following bearing capacity factors are obtained:

$$N_q = 37$$

$$N_\gamma = 42$$

$$q_{\text{ult}} = 1.2cN_c \overset{0}{} + (128)(5)(37) + (0.4)(128)(6)(42) = 36{,}600 \text{ psf, or } 36.6 \text{ ksf}$$

$$q_{\text{actual}} = \frac{Q}{A} = \frac{300}{6 \times 6} = 8.33 \text{ ksf}$$

Factor of safety against bearing capacity failure (collapse)

$$= \frac{q_{\text{ult}}}{q_{\text{actual}}} = \frac{36.6}{8.33} = 4.4 > 3.0 \quad \text{O.K.}$$

Effect of Water Table on Bearing Capacity [6]

Up to this point in this discussion of bearing capacity, it has been assumed that the water table was well below the footings and thus did not affect the soil's bearing capacity. This is not always the case, however. Depending on where the water table is located, two terms in Eqs. (6-1) through (6-3)—the γBN_γ (or γRN_γ) term and the $\gamma D_f N_q$ term—may require modification.

If the water table is at or above the base of the footing, the unit weight of the soil (γ) in the γBN_γ (or γRN_γ) terms should be the submerged unit weight (unit weight of soil minus the unit weight of water). If the water table is at a distance B or more below the base of the footing (see Fig. 6-11), the water table is assumed to have no effect, and the full unit weight of the soil should be used. If the water table is below the base of the footing but less than a distance B below the base, a linear interpolated value of effective unit weight should be used in the γBN_γ (or γRN_γ) terms. (That is, the effective unit weight is considered to vary linearly from the value of the submerged unit weight at the base of the footing to the full unit weight of the soil at a distance B below the base of the footing.)

If the water table is at the ground surface, the unit weight of the soil (γ) in the $\gamma D_f N_q$ terms of Eqs. (6-1) through (6-3) should be the submerged unit weight. If the water table is at the base of the footing, the full unit weight

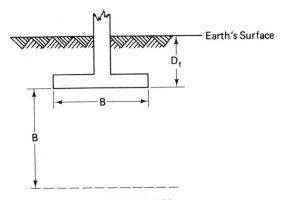

FIGURE 6-11

of the soil should be used in these terms. If the water table is between the base of the footing and the ground surface, a linear interpolated value of effective unit weight should be used in the $\gamma D_f N_q$ terms. (That is, the effective unit weight is considered to vary linearly from the value of the submerged unit weight at the ground surface to the full unit weight at the base of the footing.)

Example 6-4 deals with a square footing in soft, loose soil with the groundwater table located at the ground surface.

EXAMPLE 6-4

Given

1. A 7-ft by 7-ft square footing is located 6 ft below the ground surface (see Fig. 6-12).

2. The groundwater table is located at the ground surface.

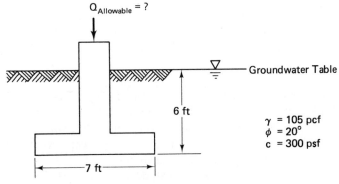

FIGURE 6-12

3. The subsoil consists of a uniform deposit of soft, loose soil. The laboratory test results are as follows:

$$\text{Angle of internal friction} = 20°$$
$$\text{Cohesion} = 300 \text{ psf}$$

The unit weight of soil $= 105$ pcf.

Required

The allowable (design) load that can be imposed onto this square footing, using a factor of safety of 3.

Solution

Since the footing is resting on soft, loose soil, Eq. (6-3) must be modified to reflect a local shear condition.

$$q_{\text{ult}} = 1.2c'N_c' + \gamma D_f N_q' + 0.4\gamma BN_\gamma'$$
$$c' = \tfrac{2}{3}c = \tfrac{2}{3} \times 300 = 200 \text{ psf}$$

With $\phi = 20°$, Fig. 6-6 gives

$$N_c' = 10$$
$$N_q' = 3$$
$$N_\gamma' = 2$$
$$B = 7 \text{ ft}$$
$$D_f = 6 \text{ ft}$$
$$\gamma = 105 - 62.4 = 42.6 \text{ pcf} \qquad \text{(below the water table, the submerged soil unit weight is used)}$$
$$q_{\text{ult}} = (1.2)(200)(10) + (42.6)(6)(3) + (0.4)(42.6)(7)(2) = 3410 \text{ psf}$$
$$q_a = \frac{3410}{3} = 1140 \text{ psf}$$

$$Q_{\text{allowable}} = q_a \times \text{area of footing} = (1140)(7)(7) = 55,900 \text{ lb} = 55.9 \text{ kips}$$

Inclined Load

If a footing is subjected to an inclined load (Fig. 6-13), the inclined load can be resolved into a vertical and a horizontal component. The vertical component can then be used for bearing capacity analysis in the same manner as described previously. After the bearing capacity has been computed by the normal procedure, it must be corrected by an R_i factor, which can be obtained from Fig. 6-14. The stability of the footing with regard to the horizontal component of the inclined load must be checked by calculating the factor of safety against sliding (see Sec. 6-6).

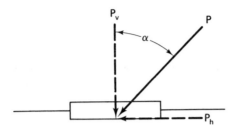

FIGURE 6-13 Footing subjected to an inclined load.

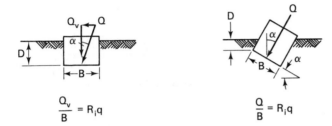

$$\frac{Q_v}{B} = R_i q \qquad\qquad \frac{Q}{B} = R_i q$$

q = Ultimate (or allowable) bearing capacity of horizontal footing under vertical load

R_i = Reduction factor; see charts below

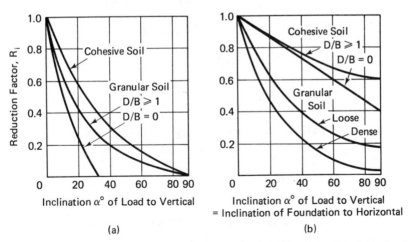

FIGURE 6-14 Inclined load reduction factors. (a) horizontal foundation [7]; (b) inclined foundation [8]. [1]

EXAMPLE 6-5

Given

A square footing (5 ft by 5 ft) is subjected to an inclined load as shown in Fig. 6-15.

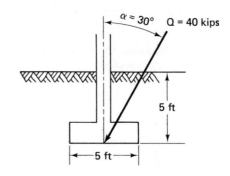

Stiff Cohesive Soil

γ = 130 pcf
q_u = 3600 psf

FIGURE 6-15

Required

Compute the factor of safety against bearing capacity failure (collapse).

Solution

For a square footing,

$$q_{ult} = 1.2cN_c + \gamma D_f N_q + 0.4\gamma B N_\gamma \qquad (6-3)$$

$$c = \frac{q_u}{2} = \frac{3600}{2} = 1800 \text{ psf}$$

$$\gamma = 130 \text{ pcf}$$

$$D_f = 5 \text{ ft}$$

$$B = 5 \text{ ft}$$

Use $c > 0$ and $\phi = 0$ analysis for cohesive soil. Fig. 6-6 gives

$$N_c = 5.14$$

$$N_q = 1.0$$

$$N_\gamma = 0$$

$$q_{ult} = (1.2)(1800)(5.14) + (130)(5)(1) + (0.4)(130)(5)(0)$$

$$= 11,800 \text{ psf} = 11.8 \text{ ksf}$$

From Fig. 6-14, with $\alpha = 30°$ and cohesive soil, the reduction factor for inclined load = 0.4.

Corrected q_{ult} for inclined load $= (0.4)(11.8) = 4.7$ ksf

$Q_v = Q \cos 30° = (40)(0.866) = 34.6$ kips

Factor of safety $= \dfrac{Q_{ult}}{Q_v} = \dfrac{(4.7)(5 \times 5)}{34.6} = 3.4$

Eccentric Loading [1]

Design of a footing is somewhat more complicated if it must support an eccentric load. Eccentric loads result from loads applied somewhere other than the centroid of the footing or from applied moments, such as those resulting at the base of a tall column from wind loads on the structure. Footings with eccentric loads may be analyzed for bearing capacity by two methods: (1) the concept of "useful width" and (2) the application of "reduction factors."

In the useful width method, only that part of the footing that is symmetrical with regard to the load is used to determine bearing capacity by the usual method, with the remainder of the footing being ignored. Thus, in Fig. 6-16, with the (eccentric) load applied at the point indicated, the shaded area is symmetrical with regard to the load, and it is used to determine bearing capacity. That area is equal to $L \times (B - 2e_b)$ in this example.

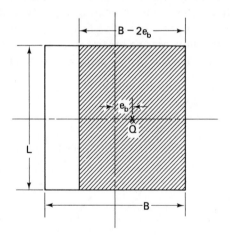

FIGURE 6-16 [1]

Upon reflection, it will be observed that this method means mathematically that the bearing capacity decreases linearly as the eccentricity (distance e_b in Fig. 6-16) increases. This linear relationship has been confirmed in the case of cohesive soils. In the case of cohesionless soils, however, a more nearly parabolic bearing capacity reduction has been determined [8]. The linear relationship for cohesive soils and the parabolic relationship for cohesionless soils are illustrated in Fig. 6-17. Since the useful width method is based on a linear bearing capacity reduction, it is recommended that this method be used only in the case of a cohesive soil.

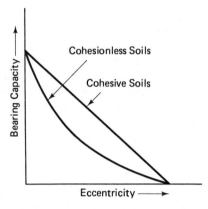

FIGURE 6-17

To use the reduction factors method, the bearing capacity is computed by the normal procedure on the assumption that the load is applied at the centroid of the footing. This value of bearing capacity is then corrected for eccentricity by multiplying by a reduction factor (R_e) obtained from Fig. 6-18.

Example 6-6 shows how bearing capacity can be calculated for an eccentric load in a cohesive soil by each of the methods described above.

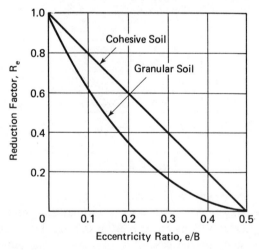

FIGURE 6-18 Eccentric load reduction factors. [1, 7]

EXAMPLE 6-6

Given

1. A 5-ft by 5-ft square footing is located 4 ft below the ground surface.
2. The footing is subjected to an eccentric load of 75 kips (see Fig. 6-19).

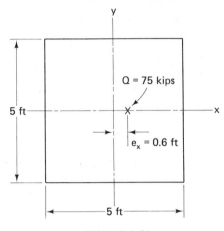

FIGURE 6-19

3. The subsoil consists of a thick deposit of cohesive soil with $q_u = 4.0$ ksf and $\gamma = 130$ pcf.
4. The water table is at very great depth and its effect on bearing capacity may be ignored.

Required

Calculate the factor of safety against bearing capacity failure (collapse) by:

1. The concept of useful width.
2. Using a reduction factor from Fig. 6-18.

Solution

1. The concept of useful width:
   From Fig. 6-20, the useful width is 3.8 ft.

$$q_{ult} = 1.2cN_c + \gamma D_f N_q + 0.4\gamma BN_\gamma \qquad (6\text{-}3)$$

$$c = \frac{q_u}{2} = \frac{4}{2} = 2 \text{ ksf}$$

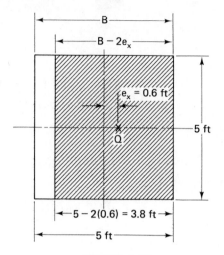

FIGURE 6-20

Use $c > 0$, $\phi = 0$ analysis for cohesive soil. From Fig. 6-6,

$$N_c = 5.14$$
$$N_q = 1.0$$
$$N_\gamma = 0$$
$$\gamma = 0.13 \text{ kcf}$$
$$B = \text{useful width} = 3.8 \text{ ft}$$
$$q_{\text{ult}} = (1.2)(2)(5.14) + (0.13)(4)(1) + (0.4)(0.13)(3.8)(0)$$
$$= 12.9 \text{ ksf}$$

$$\text{Factor of safety} = \frac{q_{\text{ult}}}{q_a} = \frac{12.9}{\left(\dfrac{75}{3.8 \times 5}\right)} = 3.27$$

2. Using a reduction factor from Fig. 6-18:

$$\text{Eccentricity ratio} = \frac{e_x}{B} = \frac{0.6}{5} = 0.12$$

For cohesive soil, Fig. 6-18 gives $R_e = 0.76$. Here q_{ult} is computed based on the actual $B = 5$ ft.

$$q_{\text{ult}} = 1.2cN_c + \gamma D_f N_q + 0.4\gamma B N_\gamma \qquad (6\text{-}3)$$
$$c = 2 \text{ ksf}$$
$$N_c = 5.14$$
$$N_q = 1.0$$
$$N_\gamma = 0$$

$$\gamma = 0.13 \text{ kcf}$$

$$B = 5 \text{ ft}$$

$$q_{ult} = (1.2)(2)(5.14) + (0.13)(4)(1) + (0.4)(0.13)(5)(0) = 12.9 \text{ ksf}$$

q_{ult} corrected for eccentricity $= q_{ult} \times R_e = (12.9)(0.76) = 9.80 \text{ ksf}$

$$\text{Factor of safety} = \frac{9.80}{\left(\dfrac{75}{5 \times 5}\right)} = 3.27$$

6-5 SIZE OF FOOTINGS

After the allowable bearing capacity of the soil has been determined, the required area of the footing is determined by dividing the footing load by the allowable bearing pressure.

The following two example problems illustrate the sizing of a footing based on allowable bearing capacity.

EXAMPLE 6-7

Given

1. A square footing rests on a uniform thick deposit of stiff clay with unconfined compressive strength (q_u) of 2.4 ksf.

2. The footing is located 4 ft below the ground surface and is to carry a total load of 250 kips (see Fig. 6-21).

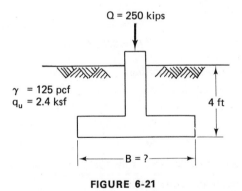

Q = 250 kips

γ = 125 pcf
q_u = 2.4 ksf

4 ft

B = ?

FIGURE 6-21

3. The unit weight of the clay is 125 pcf.

4. The groundwater is at very great depth.

Required

Determine the required square footing dimension. Use a factor of safety of 3.

Solution

Since the supporting stratum is stiff clay, a condition of general shear governs this case.

$$q_{ult} = 1.2cN_c + \gamma D_f N_q + 0.4\gamma B N_\gamma \qquad (6-3)$$

$$B = ? \text{ ft}$$

$$c = \frac{q_u}{2} = \frac{2.4}{2} = 1.2 \text{ ksf}$$

Assuming $\phi = 0$, from Fig. 6-6,

$$N_c = 5.14$$
$$N_q = 1.0$$
$$N_\gamma = 0$$
$$\gamma = 0.125 \text{ kcf}$$
$$D_f = 4 \text{ ft}$$
$$q_{ult} = (1.2)(1.2)(5.14) + (0.125)(4)(1.0) + (0.4)(0.125)(B)(0)$$
$$= 7.90 \text{ ksf}$$

$$q_a = \frac{q_{ult}}{\text{F.S.}} = \frac{7.90}{3} = 2.63 \text{ ksf}$$

The required footing area $= \dfrac{250}{2.63} = 95 \text{ ft}^2$

Therefore,

$$B^2 = 95 \text{ ft}^2$$
$$B = 9.75 \text{ ft}$$

Use a 10-ft by 10-ft square footing.

EXAMPLE 6-8

Given

1. A uniform soil deposit has the following properties:

   Unit weight of soil $= 130$ pcf

   $\phi = 30°$

   $c = 800$ psf

2. A proposed footing is to be located 5 ft below the ground surface and is to carry a total load of 600 kips (see Fig. 6-22).

3. The groundwater table is at a great depth and its effect may be ignored.

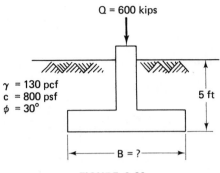

FIGURE 6-22

Required

Determine the required dimension of a square footing to carry the proposed total load of 600 kips using general shear condition and a factor of safety of 3.

Solution

$$q_{ult} = 1.2cN_c + \gamma D_f N_q + 0.4\gamma BN_\gamma \qquad (6\text{-}3)$$

$$c = 800 \text{ psf}$$

$$\gamma = 130 \text{ pcf}$$

$$D_f = 5 \text{ ft}$$

$$\phi = 30°$$

From Fig. 6-6,

$$N_c = 30$$

$$N_q = 18$$

$$N_\gamma = 17$$

First trial

Assume that $B = 10$ ft.

$$q_{ult} = (1.2)(800)(30) + (130)(5)(18) + (0.4)(130)(10)(17) = 49,300 \text{ psf}$$

$$q_a = \frac{q_{ult}}{\text{F.S.}} = \frac{49,300}{3} = 16,400 \text{ psf}$$

The required footing area $= \dfrac{600,000}{16,400} = 36.6 \text{ ft}^2$

$$B^2 = 36.6 \text{ ft}^2$$

$$B = 6.05 \text{ ft}$$

Second trial

Assume that $B = 6$ ft.

$$q_{ult} = (1.2)(800)(30) + (130)(5)(18) + (0.4)(130)(6)(17) = 45,800 \text{ psf}$$

$$q_a = \frac{45,800}{3} = 15,300 \text{ psf}$$

The required footing area $= \frac{600,000}{15,300} = 39.2 \text{ ft}^2$

$$B^2 = 39.2 \text{ ft}^2$$

$$B = 6.26 \text{ ft}$$

Use a 6.5-ft by 6.5-ft square footing.

A footing sized in the manner just described and illustrated should then be checked for settlement (see Chap. 4). If settlement is excessive (see Sec. 6-7), the size of footing should be revised.

6-6 CONTACT PRESSURE

The pressure acting between the base of a footing and the soil below is referred to as the *contact pressure*. A knowledge of contact pressure and the associated shear and moment distribution is important in footing design.

Contact pressure can be computed using the flexural formula [1]:

$$q = \frac{Q}{A} \pm \frac{M_x x}{I_y} \pm \frac{M_y y}{I_x} \tag{6-4}$$

where q = contact pressure, ksf

Q = total axial vertical load, kips

A = area of footing, ft^2

M_x, M_y = total moment parallel to respective x and y axes, ft-kips

I_x, I_y = moment of inertia about respective x and y axes, ft^4

x, y = distance from centroid to the point at which the contact pressure is computed along respective x and y axes, ft

In the special case where moments about both the x and y axes are zero, the contact pressure is simply equal to the total vertical load divided by the area of the footing. In theory, the contact pressure in this special case is uniform; but in practice, it varies somewhat because of distortion settlement. It is generally assumed to be uniform, however, for design purposes.

The use of the flexural formula to determine contact pressure is illustrated by the following example problems. Example 6-9 illustrates the computation of contact pressure when there is no moment applied to either the x or the y axis. Examples 6-10 and 6-11 illustrate the computation when there is moment applied to one axis.

EXAMPLE 6-9

Given

     1.   A 5-ft by 5-ft square footing as shown in Fig. 6-23.

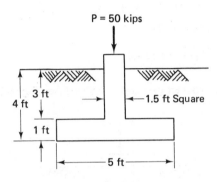

FIGURE 6-23

     2.   Centric column load on the footing = 50 kips.

     3.   Unit weight of soil = 120 pcf.

     4.   Unit weight of concrete = 150 pcf.

     5.   Ultimate bearing capacity = 9000 psf.

Required

     1.   Soil contact pressure.

     2.   Factor of safety against bearing capacity failure (collapse).

Solution

     1.   *Soil contact pressure:*

$$q = \frac{Q}{A} \pm \frac{M_x x}{I_y} \pm \frac{M_y y}{I_x} \tag{6-4}$$

Since the column load is imposed on the centroid of the footing, $M_x = 0$ and $M_y = 0$.

Q = total axial vertical load on the footing base

= column load + weight of base pad of footing + weight of pedestal of footing + weight of backfill soil

Column load = 50 kips   (given)

Weight of base pad of footing = $5 \times 5 \times 1 \times 0.15 = 3.75$ kips

Weight of pedestal of footing = $1.5 \times 1.5 \times 3 \times 0.15$
$$= 1.01 \text{ kips}$$

Weight of backfill soil = $[(5 \times 5) - (1.5 \times 1.5)] \times 3 \times 0.12$
$$= 8.19 \text{ kips}$$

$Q = 50 + 3.75 + 1.01 + 8.19 = 62.95$ kips

$A = (5)(5) = 25$ ft²

$q = \dfrac{62.95}{25} = 2.52$ ksf

Thus, soil contact pressure = 2.52 ksf (see Fig. 6-24).

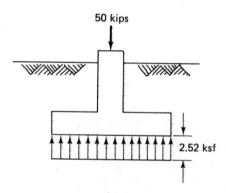

50 kips

2.52 ksf

FIGURE 6-24

2. *Factor of safety against bearing capacity failure (collapse):*
If $q_{ult} = 9000$ psf,

$$\text{F.S.} = \frac{q_{ult}}{q} = \frac{9.0}{2.52} = 3.57$$

EXAMPLE 6-10

Given

1. A 6-ft by 6-ft square column footing as shown in Fig. 6-25.
2. The base of the column is hinged.

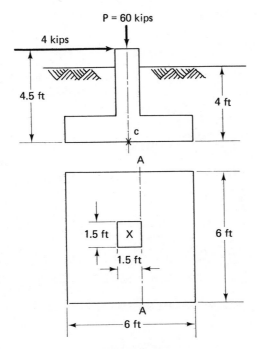

FIGURE 6-25

3. Load on the footing from the column $(P) = 60$ kips.

    Weight of concrete footing including pedestal and base pad $(W_1) = 9.3$ kips.
    Weight of backfill soil $(W_2) = 11.2$ kips.

4. Horizontal load acting on the base of the column $= 4$ kips.

5. Allowable bearing capacity of the supporting soil $= 3.0$ ksf.

Required

1. Contact pressure and soil pressure diagram.

2. Shear and moment at section A–A (see Fig. 6-25).

3. Factor of safety against sliding if the coefficient of friction between footing base and supporting soil is 0.40.

4. Factor of safety against overturning.

Solution

1. *Contact pressure and soil pressure diagram:*

$$q = \frac{Q}{A} \pm \frac{M_x x}{I_y} \pm \frac{M_y y}{I_x} \qquad (6\text{-}4)$$

Q = total vertical load = $P + W_1 + W_2 = 60 + 9.3 + 11.2$

  = 80.5 kips

$A = 6 \times 6 = 36$ ft²

$M_x = 4 \times 4.5 = 18$ ft-kips   (take moment at point C; see Fig. 6-25)

$$x = \frac{6}{2} = 3 \text{ ft}$$

$$I_y = \frac{(6)(6)^3}{12} = 108 \text{ ft}^4$$

$$M_y = 0$$

$$\frac{M_y y}{I_x} = 0$$

$$q = \frac{80.5}{36} \pm \frac{(18)(3)}{108} = 2.24 \pm 0.50$$

$q_{\text{right}} = 2.24 + 0.50 = 2.74$ ksf < 3.0 ksf $\therefore$ O.K.

$q_{\text{left}} = 2.24 - 0.50 = 1.74$ ksf < 3.0 ksf $\therefore$ O.K.

The pressure diagram is shown in Fig. 6-26.

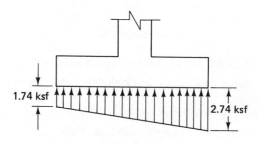

FIGURE 6-26

2. *Shear and moment at section A–A:*

From Fig. 6-27, $\triangle FDG$ and $\triangle EDH$ are similar triangles. Therefore,

$$\frac{DE}{DF} = \frac{EH}{FG}$$

$$DF = 2.74 - 1.74 = 1.0 \text{ ksf}$$

$$EH = \frac{6}{2} - \frac{1.5}{2} = 2.25 \text{ ft}   \text{ (see Figs. 6-25 and 6-27)}$$

$$FG = 6 \text{ ft}$$

$$\frac{DE}{1.0} = \frac{2.25}{6}$$

$$DE = 0.375 \text{ ksf}$$

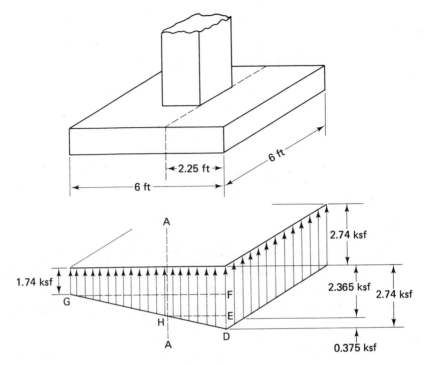

FIGURE 6-27

Shear at A–A = (2.25)(2.365)(6) + (1/2)(2.25)(0.375)(6) (see Fig.
⤸6 ft length of footing⤸ 6-27)

= 31.9 + 2.53 = 34.4 kips

Moment at A–A = $(31.9)\left(\dfrac{2.25}{2}\right)$ + (2.53)(2/3 × 2.25) = 39.7 ft-kips

3. *Factor of safety against sliding:*

Factor of safety against sliding

$$= \frac{\text{total vertical load times coefficient of friction between base and soil}}{\sum \text{horizontal forces}}$$

$$= \frac{(60 + 9.3 + 11.2)(0.4)}{4} = 8.05$$

4. *Factor of safety against overturning:*

See Fig. 6-28. By taking moments at point *K*, the factor of safety against overturning can be computed as follows:

$$\text{F.S.} = \frac{\text{moment to resist turning}}{\text{turning moment}} = \frac{(80.5)(6/2)}{(4)(4.5)} = 13.4$$

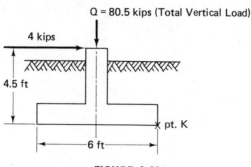

FIGURE 6-28

EXAMPLE   6-11

Given

1. A 7.5-ft by 10-ft rectangular column footing as shown in Fig. 6-29.
2. The base of the column is fixed into the foundation.
3. Load on the footing from the column (P) = 50 kips.
   Weight of the concrete footing and weight of the backfill soil (w) = 25 kips.

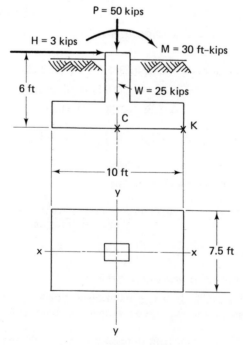

FIGURE 6-29

Horizontal load acting on the base of the column $(H) = 3$ kips.
Moment acting on the foundation $(M) = 30$ ft-kips.

4. Allowable bearing capacity of the soil $= 2$ ksf.

Required

1. Contact pressure and soil pressure diagram.
2. Factor of safety against overturning.

Solution

1. *Contact pressure and soil pressure diagram:*

$$q = \frac{Q}{A} \pm \frac{M_x x}{I_y} \pm \frac{M_y y}{I_x} \qquad (6\text{-}4)$$

$$Q = 50 + 25 = 75 \text{ kips}$$

$$A = 7.5 \times 10 = 75 \text{ ft}^2$$

$$M_x = (3)(6) + 30 = 48 \text{ ft-kips} \quad \text{(take moments at point } C;$$
$$\text{see Fig. 6-29)}$$

$$x = \frac{10}{2} = 5 \text{ ft}$$

$$I_y = \frac{(7.5)(10)^3}{12} = 625 \text{ ft}^4$$

$$M_y = 0$$

$$q = \frac{75}{75} \pm \frac{(48)(5)}{625} = 1.0 \pm 0.38$$

$$q_{\text{right}} = 1.38 \text{ ksf} < 2 \text{ ksf} \qquad \therefore \text{ O.K.}$$

$$q_{\text{left}} = 0.62 \text{ ksf} < 2 \text{ ksf} \qquad \therefore \text{ O.K.}$$

The pressure diagram is shown in Fig. 6-30.

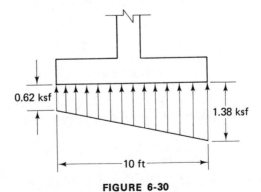

FIGURE 6-30

2. *Factor of safety against overturning:*

By taking moments at point K (Fig. 6-29),

$$\text{F.S.} = \frac{\text{moment to resist turning}}{\text{turning moment}} = \frac{(50 + 25)(10/2)}{(3)(6) + (30)} = 7.8$$

Under certain conditions, such as very large applied moments, Eq. (6-4) may give a negative value for the contact pressure. This implies tension between the footing and the soil. Soil cannot furnish any tensile resistance; hence, the flexural formula is not applicable in this situation. Instead, the contact pressure may be calculated according to the basic equations of statics in the following manner.

Referring to Fig. 6-31, by summing all forces in the vertical direction and

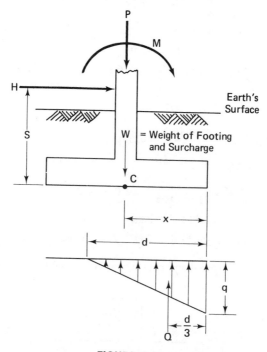

FIGURE 6-31

setting this sum equal to zero, and by summing all moments about point C and setting this sum equal to zero, the following two equations are obtained.

$$\sum V = 0 \uparrow_+ \qquad \left(\frac{q}{2}\right)(d)(L) - P - W = 0$$

$$\sum M_c = 0 \curvearrowright_+ \qquad M + (H)(S) - \left(\frac{q}{2}\right)(d)(L)\left(x - \frac{d}{3}\right) = 0$$

Since all terms except q and d in these two equations are known, the two simultaneous equations may be solved to determine q and d. With q and d both known, the soil pressure diagram may be drawn. This technique is illustrated by Example 6-12.

EXAMPLE 6-12

Given

A rectangular footing 5 ft by 7.5 ft loaded as shown in Fig. 6-32.

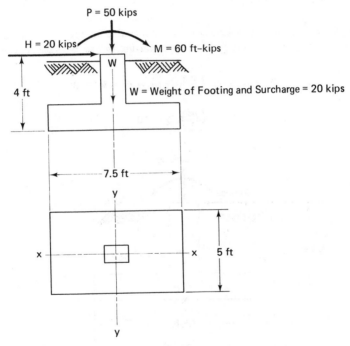

FIGURE 6-32

Required

Compute contact pressure and draw the soil pressure diagram.

Solution

By the flexural formula,

$$q = \frac{Q}{A} \pm \frac{M_x x}{I_y} \pm \frac{M_y y}{I_x} \qquad (6\text{-}4)$$

$$Q = 50 + 20 = 70 \text{ kips}$$

$$A = 5 \times 7.5 = 37.5 \text{ ft}^2$$

$$M_x = (4)(20) + 60 = 140 \text{ ft-kips} \quad \text{(take moments at point } C; \\ \text{see Fig. 6-32)}$$

$$x = \frac{7.5}{2} = 3.75 \text{ ft}$$

$$I_y = \frac{(5)(7.5)^3}{12} = 176 \text{ ft}^4$$

$$q = \frac{70}{37.5} \pm \frac{(140)(3.75)}{176} = 1.87 \pm 2.98$$

$$\left.\begin{array}{l} q_{\text{right}} = +4.85 \text{ ksf} \\ q_{\text{left}} = -1.11 \text{ ksf} \end{array}\right\} \quad q \text{ has a negative answer}$$

Therefore, the flexural formula is not applicable in this case. Solve this problem by $\sum V = 0$ and $\sum M_c = 0$. Referring to Fig. 6-32 and Fig. 6-33,

$$\sum V = 0 \qquad \left(\frac{q}{2}\right)(d)(L) - P - W = 0$$

$$\left(\frac{qd}{2}\right)(5) = 70 \tag{A}$$

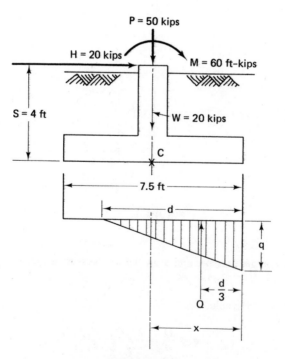

FIGURE 6-33

$$\sum M_c = 0 \qquad M + (H)(S) - \left(\frac{q}{2}\right)(d)(L)\left(x - \frac{d}{3}\right) = 0$$

$$60 + (20)(4) - (70)\left(\frac{7.5}{2} - \frac{d}{3}\right) = 0 \qquad \text{(B)}$$

[Note that $(qd/2)(L)$ equals 70, from (A).]
From (B),

$$60 + 80 - 262.5 + \frac{70}{3}d = 0$$

$$d = 5.25 \text{ ft}$$

Substitute $d = 5.25$ ft into (A):

$$\left(\frac{q}{2}\right)(5.25)(5) = 70$$

$$q = 5.33 \text{ ksf}$$

The pressure diagram is shown in Fig. 6-34.

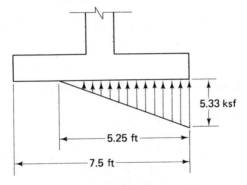

FIGURE 6-34

6-7 TOTAL AND DIFFERENTIAL SETTLEMENT

Previous material in this chapter has dealt primarily with bearing capacity analysis and prevention of bearing capacity failure (collapse) of footings. Footings may also fail as a result of excessive settlement; thus, after the size of footing has been determined by bearing capacity, footing settlement should be calculated and the design revised if the calculated settlement is considered to be excessive.

The calculation of settlement has already been covered (Chap. 4). Maximum permissible settlement depends primarily on the nature of the superstructure. Some suggested maximum permissible settlement values are given in Table 6-1.

TABLE 6-1   Maximum permissible settlement [9].

Limiting Factor or Type of Structure	Maximum Allowable Settlement	
	Differential[1]	Total (in.)
Drainage of floors	0.01–0.02L	6–12
Stacking, warehouse lift trucks	0.01L	6
Tilting of smokestacks, silos	0.004B	3–12
Framed structure, simple	0.005L	2–4
Framed structure, continuous	0.002L	1–2
Framed structure with diagonals	0.0015L	1–2
Reinforced concrete structure	0.002–0.004L	1–3
Brick walls, one-story	0.001–0.002L	1–2
Brick walls, high	0.0005–0.001L	1
Cracking of panel walls	0.003L	1–2
Cracking of plaster	0.001L	1
Machine operation, noncritical	0.003L	1–2
Crane rails	0.003L	
Machines, critical	0.0002L	

[1] L is the distance between adjacent columns; B is the width of base.

6-8   STRUCTURAL DESIGN OF FOOTINGS

As has been noted in Sec. 6-5, the required base area of a footing may be determined by dividing the column load by the allowable bearing capacity. Determining the thickness and shape of the footing, the amount and location of reinforcing steel, and other actual structural design of footings is, however, ultimately the responsibility of the structural engineer.

In general, it is the responsibility of the soils engineer or technologist to furnish the contact pressure diagram and the shear as well as the moment at a section (in the footing) at the face of the column, pedestal, or wall. This was demonstrated in Example 6-10 when the contact pressure diagram, the shear, and the moment at section A–A were determined. From this information, the structural engineer can do the actual structural design of the footing.

6-9   PROBLEMS

6-1   A strip of wall footing 3 ft wide is located 3.5 ft below the ground surface. The supporting soil has a unit weight of 125 pcf. The results of the laboratory tests on the soil samples indicate that the cohesion and the angle of internal friction of the supporting soil are 1200 psf and 25°, respectively. Groundwater was not encountered during the subsurface soil exploration. Determine the allowable bearing capacity, using a factor of safety of 3.

6-2 A square footing with a size of 10 ft by 10 ft is located 8 ft below the ground surface. The subsoil consists of a thick deposit of stiff cohesive soil with unconfined compressive strength (q_u) equal to 3600 psf. The unit weight of this cohesive soil is 128 pcf. Compute the ultimate bearing capacity.

6-3 A footing 8 ft by 8 ft is buried 6 ft below the ground surface in a dense cohesionless soil. The results of laboratory and field tests of the supporting soil indicate that the unit weight of soil is 130 pcf and the average corrected standard penetration test value beneath the footing is 37. Compute the allowable (design) load that can be imposed onto this square footing, using a factor of safety of 3.

6-4 A square footing with a size of 8 ft by 8 ft is to carry a total load of 40 kips. The depth of footing is 5 ft below the ground surface and groundwater is located at the ground surface. The subsoil consists of a uniform deposit of soft clay, and cohesion of the soil is 500 psf. The unit weight of the soil equals 110 pcf. Compute the factor of safety against bearing capacity failure.

6-5 A square footing will be constructed on a uniform thick deposit of clay with unconfined compressive strength (q_u) of 3 ksf. The footing will be located 5 ft below the ground surface and is designed to carry a total load of 300 kips. The unit weight of the supporting soil is 128 pcf. No groundwater was encountered during soil exploration. Considering general shear, determine the square footing dimension, using a factor of safety of 3.

6-6 A proposed square footing carrying a total load of 500 kips is to be constructed on a uniform thick deposit of dense cohesionless soil. The properties of the sand deposit are as follows. Unit weight of soil is 135 pcf and angle of internal friction is 38°. The depth of footing is to be 5 ft. Determine the dimension of this proposed square footing, using a factor of safety of 3.

6-7 Compute and draw the soil pressure diagrams for the footing shown in Fig. 6-35 for the following loads:

(a) $P = 70$ kips and $H = 20$ kips

(b) $P = 70$ kips and $H = 10$ kips

6-8 Considering general shear, compute the safety factor against a bearing capacity failure for each of the two loadings in Problem 6-7 if the bearing soil is:

(a) Cohesionless   (b) Cohesive
    $\phi = 30°$ $\phi = 0°$
    $\gamma = 110$ pcf $\gamma = 110$ pcf
    $c = 0$ $c = 3000$ psf

Note: The groundwater is 10 ft below the base of the footing.

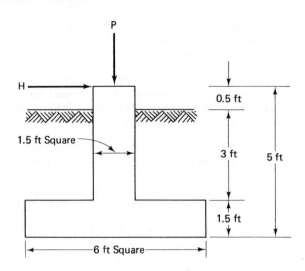

Concrete Unit Weight = 150 pcf
Soil Unit Weight = 110 pcf

FIGURE 6-35

6-9 Same as Problem 6-7 except that the groundwater is located at the ground surface.

6-10 A footing is shown in Fig. 6-36. Vertical load, including column load, surcharge weight, and weight of footing, is 120 kips. Horizontal load is 10 kips and a moment of 50 ft-kips (clockwise) is also imposed on the foundation.

(a) Compute the soil contact pressure and draw the soil contact pressure diagram.

(b) Compute the shear on section a–a (Fig. 6-36).

(c) Compute moment on section a–a (Fig. 6-36).

(d) Compute the factor of safety against overturning.

(e) Compute the factor of safety against sliding, if the coefficient of friction between the soil and the base of the footing is 0.60.

(f) Compute the factor of safety against bearing capacity failure if the ultimate bearing capacity of the foundation soil supporting the subject footing is 5.4 tons/ft².

6-11 A 6-ft by 6-ft square footing is buried 5 ft below the ground surface. The footing is subjected to an eccentric load of 200 kips. The eccentricity of the 200-kip load (e_x) is 0.8 ft. The supporting soil consists of cohesionless soil with $\phi = 38°$, $c = 0$, and $\gamma = 135$ pcf. Calculate the factor of safety against bearing capacity failure (collapse) by a reduction factor from Fig. 6-18.

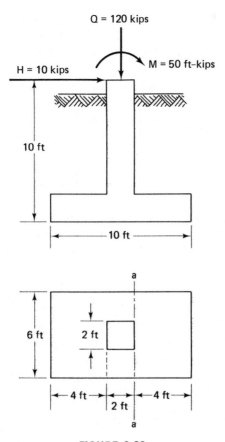

FIGURE 6-36

References

[1] WAYNE C. TENG, *Foundation Design*, Prentice-Hall, Inc., Englewood Cliffs, N.J., 1962.

[2] G. A. LEONARDS, ED., *Foundation Engineering*, McGraw-Hill Book Company, New York, 1962.

[3] KARL TERZAGHI AND RALPH B. PECK, *Soil Mechanics in Engineering Practice*, John Wiley & Sons, Inc., New York, 1967. Copyright © 1967, by John Wiley & Sons, Inc. Reprinted by permission of John Wiley & Sons, Inc.

[4] G. G. MEYERHOF, "Influence of Roughness Base and Groundwater Conditions on the Ultimate Bearing Capacity of Foundations," *Geotechnique*, **5**, 227–242 (1955).

[5] RALPH B. PECK, WALTER E. HANSEN, AND THOMAS H. THORNBURN, *Foundation Engineering*, 2nd ed., John Wiley & Sons, Inc., New York, 1974. Copyright © 1974, by John Wiley & Sons, Inc. Reprinted by permission of John Wiley & Sons, Inc.

[6] DAVID F. MCCARTHY, *Essentials of Soil Mechanics and Foundations*, Reston Publishing Company, Inc., Reston, Va., 1977.

[7] AREA, *Manual of Recommended Practice*, Construction and Maintenance Section, Engineering Division, Association of American Railroads, Chicago, 1958.

[8] G. G. MEYERHOF, "The Bearing Capacity of Foundations under Eccentric and Inclined Loads," *Proc. 3rd Int. Conf. Soil Mech. Found. Eng.*, *Switzerland*, **1**, 440–45 (1953).

[9] MERLIN G. SPANGLER AND RICHARD L. HANDY, *Soil Engineering*, 3rd ed., Intext Educational Publishers, New York, 1973. Copyright 1951, © 1960, 1973 by Harper & Row, Publishers, Inc. Reprinted by permission of the publisher.

7
Pile Foundations

7-1 INTRODUCTION

Chapter 6 covered shallow foundations. Sometimes, however, the soil upon which a structure is to be built is of such poor quality that a shallow foundation would be subject to bearing capacity failure and/or excessive settlement. In such cases, *pile foundations* may be used to support the structure (i.e., to transmit the load of the structure to firmer soil, or rock, at greater depth below the structure).

A pile foundation is a relatively long and slender member that is forced or driven into the soil or it may be poured in place. If the pile is driven until it rests on a hard, impenetrable layer of soil or rock, the load of the structure is primarily transmitted axially through the pile to this layer. Such a pile is called an *end-bearing* pile. In the case of an end-bearing pile, care must be exercised to ensure that the hard, impenetrable layer is adequate to support the load. If the pile cannot be driven to a hard stratum of soil or rock (e.g., such a stratum is located too far below the ground surface), the load of the structure must be borne primarily by skin friction or adhesion between the surface of the pile and adjacent soil. Such a pile is called a *friction* pile.

In addition to simply supporting the load of a structure, piles may perform other functions, such as densifying loose cohesionless soils, resisting horizontal loads, anchoring down structures subject to uplift, and so on. The emphasis in this book, however, is on piles that support the load of a structure.

7-2 TYPES OF PILES

Piles may be classified according to the types of materials of which they are made. Virtually all piles are made of timber, concrete, or steel (or a combination of these). Each of these is discussed in general terms in this section.

Timber piles have been used for centuries and are still widely used. They are made relatively easily by delimbing tall, straight tree trunks. They generally make economical pile foundations. Timber piles have certain disadvantages, however. They have less capacity to carry load than do concrete or steel piles. Also, the length of a timber pile is limited by the height of tree available. The length is generally limited to around 60 ft, although longer ones are available in some locales. Timber piles may be damaged in the pile-driving process. In addition, they are subject to decay and attack by insects. This generally is not a problem if the pile is both in soil and always below the water table; if above the water table, the pile can be treated chemically to increase its life.

Concrete piles may be either precast or cast-in-place. Precast concrete piles may be manufactured with square or octagonal or other cross-section shape. They may be made of uniform cross section (with a pointed tip), or they may be tapered. Precast piles may be made of prestressed concrete. The main disadvantages of precast concrete piles have to do with problems of manufacturing and handling of the piles (space required, time required for curing, heavy equipment required in handling and transporting, etc.).

Cast-in-place concrete piles may be *cased* or *uncased*. The cased type can be made by driving a shell containing a core into the soil, removing the core, and filling the shell with concrete. The uncased type can be made in a similar manner, except that the shell is withdrawn as the concrete is poured. Cast-in-place concrete piles have several advantages over precast ones. One is that, since the concrete is poured in place, damage due to pile driving is eliminated. Also, the length of the pile is known at the time the concrete is poured. (In the case of the precast pile, the exact length of pile to be cast must be known initially. If a given pile turns out to be too long or too short, extra cost is involved in cutting off the extra length or adding to it.)

Concrete piles generally have a somewhat larger capacity to carry load than do timber piles. They are generally not very susceptible to deterioration, except possibly by seawater and strong chemicals.

Steel piles are usually either pipe-shaped or H-sections. Pipe-shaped steel piles may be filled with concrete after being driven. H-shaped steel piles are strong and capable of being driven to great depths through stiff layers. Steel piles are subject to damage by corrosion. They generally have a somewhat larger capacity to carry load than do timber piles or concrete piles.

Table 7-1 gives some customary design loads for different types of piles.

TABLE 7-1   Customary design loads for piles [1].

Type of Pile	Allowable Load (tons)
Wood	15–30
Composite	20–30
Cast-in-place concrete	30–50
Precast reinforced concrete	30–50
Steel pipe, concrete filled	40–60
Steel H-section	30–60

7-3 LENGTH OF PILES

In the case of end-bearing piles, the required length of pile can be determined fairly accurately, as it is the distance from the structure being supported by the pile to the hard, impenetrable layer of soil or rock on which the pile rests. This distance is established from soil boring tests.

In the case of friction piles, the required length of pile must be determined indirectly. The pile must be driven to such a depth that adequate lateral surface area of the pile is in contact with the soil in order that sufficient skin friction can be developed.

Table 7-2 gives available lengths of various types of piles.

7-4 PILE CAPACITY—THE CAPACITY OF SINGLE PILES

The capacity of single piles may be evaluated by the structural strength of the pile and by the supporting strength of the soil.

Pile Capacity as Evaluated by the Structural Strength of the Pile

Obviously, a pile must be strong enough structurally to "carry" the load imposed upon it. The structural strength of a pile depends on the size and shape of the pile as well as the type of material of which it is made.

Allowable structural strengths of different types of pile are specified by a number of different building codes. Table 7-3 summarizes allowable stress in various types of pile according to several different codes.

TABLE 7-2 Available lengths of various pile types [2].

Pile Type	Comment, Available Maximum Length
Timber	Depends on wood (tree) type. Lengths in the 50- to 60-ft range are usually available in most areas; lengths to about 75 ft are available but in limited quantity; lengths up to the 100-ft range are possible but very limited.
Steel H and pipe	Unlimited length; "short" sections are driven and additional sections are field-welded to obtain a desired total length.
Steel shell, cast-in-place	Typically to between 100 and 125 ft, depending on shell type and manufacturer-contractor.
Precast concrete	Solid, small cross-section piles usually extend into the 50- to 60-ft length, depending on cross-sectional shape, dimensions, and manufacturer. Large-diameter cylinder piles can extend to about 200 ft long.
Drilled shaft, cast-in-place concrete	Usually in the 50- to 75-ft range, depending on contractor equipment.
Bulb-type, cast-in-place concrete	Up to about 100 ft.
Composite	Related to available lengths of material in the different sections. If steel and thin-shell cast-in-place concrete are used, the length can be unlimited; if timber and thin-shell cast-in-place concrete are used, lengths can be on the order of 150 ft.

Pile Capacity as Evaluated by the Supporting Strength of the Soil

In addition to structural considerations of the strength of the pile itself, pile capacity is limited by the supporting strength of the soil. As mentioned in the introduction to this chapter (Sec. 7-1), the load carried by a pile is ultimately borne by either or both of two ways. The load is transmitted to the soil surrounding the pile by friction or adhesion between the sides of the pile and the soil, and/or the load is transmitted to the soil just below the base of the pile. This can be expressed in equation form as

$$Q_{\text{ultimate}} = Q_{\text{friction}} + Q_{\text{tip}} \tag{7-1}$$

where Q_{ultimate} = ultimate (at failure) bearing capacity of a single pile

TABLE 7-3 Allowable stress in piles [3].

Type of Pile	New York City Codes 1948	Boston City Codes 1958	Uniform Building Code 1961	AASHTO 1957	AREA 1951
Timber	600–800 psi 20 tons (6-in. ϕ tip) 25 tons (8-in. ϕ tip)	16–30 tons depending on species	60% basic stress for clear material but not exceeding 1000 psi	18 tons (10 in. ϕ) 20 tons (12 in. ϕ) 24 tons (14 in. ϕ) 28 tons (16 in. ϕ)	
Concrete[1]	0.25fc' but not exceeding 1000 psi	0.25fc' but not exceeding 900 psi	0.225fc'	20 tons (10 in. ϕ) 24 tons (12 in. ϕ) 28 tons (14 in. ϕ) 32 tons (16 in. ϕ) 40 tons (20 in. ϕ) 50 tons (24 in. ϕ)	0.25fc' (friction piles) 0.20fc' (point-bearing piles)
Steel	9000 psi (uncased) 12,000 psi (cased)	8000 psi (concrete filled pipe piles) 7000 psi (H-piles) 15,000 psi (core)	9000 psi	6000 psi	12,000[2] psi (if splice sleeves are welded) 9000[2] psi (if splice sleeves are not welded)

[1]Note: fc' = concrete cylinder strength at 28 days.
[2]Deduct $\frac{1}{16}$ in. for corrosion.

Q_{friction} = bearing capacity furnished by friction or adhesion between the sides of the pile and the soil

Q_{tip} = bearing capacity furnished by the soil just beneath the base of the pile

The term Q_{friction} in Eq. (7-1) can be evaluated by multiplying the unit skin friction or adhesion between the sides of the pile and the soil (f) by the surface area of the pile ($A_{surface}$). The term Q_{tip} in Eq. (7-1) can be evaluated by multiplying the ultimate bearing capacity of the soil at the tip of the pile (q) by the area of the tip of the pile (A_{tip}). Hence, Eq. (7-1) can be expressed as

$$Q_{ultimate} = f \cdot A_{surface} + q \cdot A_{tip} \qquad (7\text{-}2)$$

In applying Eq. (7-2), the two area terms ($A_{surface}$ and A_{tip}) are often expressed in square feet, and f and q are expressed in pounds per square foot. Thus, $Q_{ultimate}$, as computed, is expressed in pounds.

It will be noted that in the case of end-bearing piles, the term Q_{tip} of Eq. (7-1) or $q \cdot A_{tip}$ of Eq. (7-2) will be predominant; whereas in the case of friction piles, the term $Q_{friction}$ of Eq. (7-1) or $f \cdot A_{surface}$ of Eq. (7-2) will be predominant.

Equations (7-1) and (7-2) are generalized and therefore applicable for all soils. The manner in which some of the terms of Eq. (7-2) are evaluated differs, however, depending on whether the pile is driven in sand or in clay. It is convenient, therefore, to consider separately piles driven in sand and piles driven in clay.

Piles driven in sand   In the case of piles driven in sand, the skin friction between the sides of the pile and the soil [$f \cdot A_{surface}$ in Eq. (7-2)] can be evaluated by multiplying the coefficient of friction between sand and pile surface (tan δ) by the total horizontal soil pressure acting on the pile. The coefficient of friction between sand and pile surface can be obtained from Table 7-4. The total horizontal soil pressure acting on the pile is a function of the effective vertical (overburden) pressure of soil adjacent to the pile. Soil pressure normally increases as depth increases. In the special case of piles driven in soil, however, it has been determined that the effective vertical (overburden) pressure of soil adjacent to a pile does not increase without

TABLE 7-4   Coefficient of friction between sand and pile materials [2].

Material	Tan δ
Concrete	0.45
Wood	0.4
Steel (smooth)	0.2
Steel (rough, rusted)	0.4
Steel (corrugated)	Use tan ϕ of sand

limit as depth increases. Instead, the effective vertical pressure increases as depth increases until a certain depth of penetration is reached. Below this depth, which is called the critical depth and denoted D_c, the effective vertical pressure remains more or less constant (or changes very little). The critical depth is dependent on the field condition of the sand and the dimension of the pile. Tests indicate that the critical depth ranges from about 10 pile diameters for loose sand to about 20 pile diameters for dense compact sand [2]. Thus, the effective vertical pressure of soil adjacent to a pile varies with depth as illustrated in Fig. 7-1.

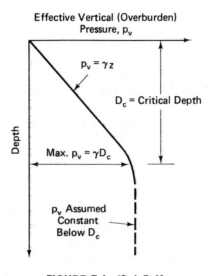

FIGURE 7-1   [2, 4, 5, 6]

The term $f \cdot A_{\text{surface}}$ of Eq. (7-2) can now be determined for a pile by multiplying the circumference of the pile by the area under the p_v versus depth curve (Fig. 7-1) by the coefficient of lateral earth pressure (K) by the coefficient of friction between sand and pile surface (tan δ). The coefficient of lateral earth pressure is assumed to vary between 0.60 and 1.25, with lower values used for silty sands and higher values used for other deposits [7].

The bearing capacity [q in Eq. (7-2)] at the pile tip can be calculated using the bearing capacity equations for cohesionless soil, which were developed by Terzaghi and Peck [1].

$$q_{\text{tip}} = \gamma D_f N_q + 0.6\gamma R N_\gamma, \qquad \text{(for circular pile)} \qquad (7\text{-}3)$$

$$q_{\text{tip}} = \gamma D_f N_q + 0.4\gamma B N_\gamma, \qquad \text{(for square pile)} \qquad (7\text{-}4)$$

where q_{tip} = bearing capacity at pile tip

           γ = unit weight of soil

           D_f = embedded length of pile

    N_γ, N_q = bearing capacity factors (see Fig. 6-6)

          R = radius of pile tip (for circular pile)

          B = width of pile tip (for square pile)

It will be noted that these equations have the same general form as the bearing capacity equations given in Chap. 6 for shallow foundations. However, as indicated previously, the magnitude of effective vertical (overburden) pressure of soil adjacent to a pile is limited below the critical depth. Thus, for design purposes, the term $\gamma D_f N_q$ in Eqs. (7-3) and (7-4) should be replaced by the term $p_v N_q$, where p_v is the effective vertical pressure adjacent to the pile at the tip of the pile (see Fig. 7-1) [2].

In most cases, driven piles are relatively small in cross section, and therefore the terms in Eqs. (7-3) and (7-4) involving R and B are small compared to the other term in the equations. Thus, for many cases, Eqs. (7-3) and (7-4) may be approximated as

$$q_{tip} = p_v N_q \qquad (7\text{-}5)$$

Thus, the term $q \cdot A_{tip}$ of Eq. (7-2) can be determined by multiplying the value of q_{tip} from Eq. (7-5) by the area of the tip of the pile.

The value of N_q is related to the angle of internal friction (ϕ) of the sand, and it should, of course, be based on the value of the angle of internal friction of the sand located in the general vicinity of where the tip of the pile will ultimately rest. The value of the angle of internal friction of the sand at this location can be determined by laboratory tests on a sample taken from the specified location or by correlation with penetration resistance tests in a boring hole (i.e., corrected SPT value) (see Figs. 2-9 and 6-7).

To summarize the method described in this section for computing pile capacity for piles driven in sand, Eq. (7-2) is used with the term $f \cdot A_{surface}$ determined by multiplying the circumference of the pile by the area under the p_v–depth curve (Fig. 7-1) by the coefficient of lateral earth pressure (K) by the coefficient of friction between sand and pile surface ($\tan \delta$) and with the term $q \cdot A_{tip}$ determined by multiplying the value of q_{tip} obtained from Eq. (7-5) by the area of the tip of the pile. The pile capacity thus determined represents the ultimate load that can be applied to the pile. In practice, it is common to apply a factor of safety of 2 to determine the (downward) design load for the pile [2].

Examples 7-1 and 7-2 illustrate the procedure for calculating pile capacity for piles driven in sand.

EXAMPLE 7-1

Given

1. A concrete pile is to be driven into a medium dense to dense sand.
2. The diameter of the pile is 12 in. and the embedded length of the pile is 25 ft.
3. Soil conditions are shown in Fig. 7-2.
4. No groundwater was encountered, and the groundwater table is not expected to rise during the life of the structure.

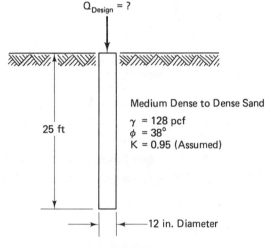

$Q_{Design} = ?$

25 ft

Medium Dense to Dense Sand

$\gamma = 128$ pcf
$\phi = 38°$
$K = 0.95$ (Assumed)

12 in. Diameter

FIGURE 7-2

Required

The approximate axial capacity of the pile if the coefficient of lateral pressure (K) is assumed to be 0.95 and the factor of safety is 2.

Solution

D_c = critical depth = 20 times the diameter of pile for dense sand = 20×1
= 20 ft (see Fig. 7-3)

From Eq. (7-2),

$$Q_{\text{ultimate}} = f \cdot A_{\text{surface}} + q \cdot A_{\text{tip}} \tag{7-2}$$

$f \cdot A_{\text{surface}}$ = total skin friction

= (circumference of pile)(area of p_v diagram)$(K)(\tan \delta)$

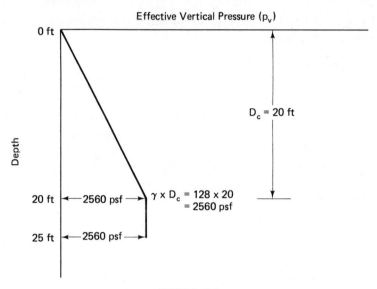

FIGURE 7-3

Circumference of pile $= \pi d = (3.14)(1) = 3.14$ ft

Area of p_v diagram $= (\frac{1}{2})(2560)(20) + (2560)(25 - 20) = 38,400$ lb/ft

$K = 0.95$ (given)

$\tan \delta = 0.45$ (see Table 7-4 for concrete pile)

$f \cdot A_{\text{surface}} = (3.14)(38,400)(0.95)(0.45) = 51,500$ lb $= 51.5$ kips

From Eq. (7-5),

$$q_{\text{tip}} = p_v N_q \qquad (7\text{-}5)$$

$$p_v = 2560 \text{ psf} \quad \text{(see Fig. 7-3)}$$

$$N_q = 50 \quad \text{(from Fig. 6-6 for } \phi = 38°)$$

$$q_{\text{tip}} = (2560)(50) = 128,000 \text{ psf}$$

$$A_{\text{tip}} = \frac{\pi d^2}{4} = \left(\frac{3.14}{4}\right)(1)^2 = 0.785 \text{ ft}^2$$

$$q \cdot A_{\text{tip}} = (128,000)(0.785) = 100,500 \text{ lb} = 100.5 \text{ kips}$$

$$Q_{\text{ultimate}} = 51.5 + 100.5 = 152 \text{ kips}$$

$$Q_{\text{design}} = \frac{Q_{\text{ultimate}}}{\text{F.S.}} = \frac{152}{2} = 76.0 \text{ kips}$$

EXAMPLE 7-2

Given

The same conditions as in Example 7-1, except that the groundwater is located 10 ft below the ground surface (see Fig. 7-4).

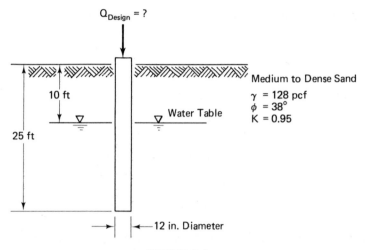

FIGURE 7-4

Required

The approximate axial capacity of the pile if K is 0.95 and a factor of safety of 2 is used.

Solution

$$D_c = 20 \text{ times the diameter of pile for dense sand}$$
$$= 20 \times 1 = 20 \text{ ft (see Fig. 7-5)}$$

$$f \cdot A_{\text{surface}} = (\text{circumference of pile})(\text{area of } p_v \text{ diagram})(K)$$
$$(\tan \delta)$$

Circumference of pile $= \pi d = (3.14)(1) = 3.14$ ft

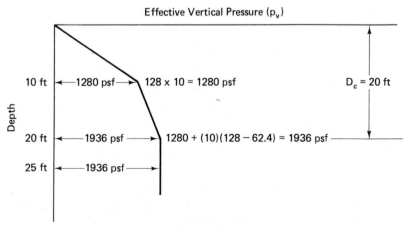

FIGURE 7-5

$$\text{Area of } p_v \text{ diagram} = (\tfrac{1}{2})(1280)(10) + (\tfrac{1}{2})(1280 + 1936)(10) + (1936)(5)$$
$$= 32{,}200 \text{ lb/ft}$$
$$K = 0.95$$
$$\tan \delta = 0.45$$
$$f \cdot A_{\text{surface}} = (3.14)(32{,}200)(0.95)(0.45) = 43{,}200 \text{ lb} = 43.2 \text{ kips}$$
$$q_{\text{tip}} = p_v N_q \tag{7-5}$$
$$q_{\text{tip}} = (1936)(50) = 96{,}800 \text{ psf}$$
$$q \cdot A_{\text{tip}} = (96{,}800)(0.785) = 76{,}000 \text{ lb} = 76.0 \text{ kips}$$
$$Q_{\text{ultimate}} = 43.2 + 76.0 = 119.2 \text{ kips}$$
$$Q_{\text{design}} = \frac{119.2}{2} = 59.6 \text{ kips}$$

Piles driven in clay   In the case of piles driven in clay, the unit adhesion between the sides of the pile and the soil [f in Eq. (7-2)] can be evaluated by multiplying the cohesion of the clay (c) by the adhesion factor (α). The adhesion factor can be determined using Fig. 7-6. The term $f \cdot A_{\text{surface}}$ of Eq. (7-2) can thus be determined by multiplying the (undisturbed) cohesion of the clay (c) by the adhesion factor (α) by the surface (skin) area of the pile (A_{surface}).

In the case of soft clays, there is a tendency for the clay to contact the pile, in which case adhesion is assumed equal to cohesion (meaning $\alpha = 1.0$). In the case of stiff clays, pile driving disturbs the surrounding soil and may cause a small open space to develop between the clay and the pile. Thus adhesion is smaller than cohesion (meaning $\alpha < 1.0$).

The bearing capacity [q in Eq. (7-2)] at the pile tip can be calculated using [2]

$$q_{\text{tip}} = c N_c \tag{7-6}$$

where q_{tip} = bearing capacity at pile tip

   c = cohesion of the clay located in the general vicinity of where the tip of the pile will ultimately rest

   N_c = bearing capacity factor and has a value of about 9 [2]

Thus, the term $q \cdot A_{\text{tip}}$ of Eq. (7-2) can be determined by multiplying the value of q_{tip} from Eq. (7-6) by the area of the tip of the pile.

To summarize the method described in this section for computing pile capacity for piles driven in clay, Eq. (7-2) is used with the term $f \cdot A_{\text{surface}}$ determined by multiplying the cohesion of the clay (c) by the adhesion factor (α) by the surface area of the pile and with the term $q \cdot A_{\text{tip}}$ determined by multiplying the value of q_{tip} obtained from Eq. (7-6) by the area of the tip of

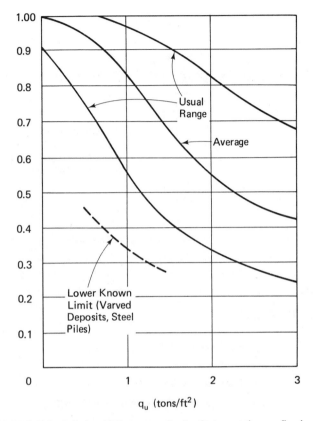

FIGURE 7-6 Relationship between adhesion factor α and unconfined compressive strength q_u. [8]

the pile. The pile capacity thus determined represents the ultimate load that can be applied to the pile. In practice, it is common to apply a factor of safety of 2 to determine the (downward) design load for the pile [2].

Examples 7-3, 7-4, and 7-5 illustrate the procedure for calculating pile capacity for piles driven in clay.

EXAMPLE 7-3

Given

1. A 12-in.-diameter concrete pile is driven at a site as shown in Fig. 7-7.
2. The embedded length of pile is 35 ft.

Required

The design capacity of the pile. Use a factor of safety of 2.

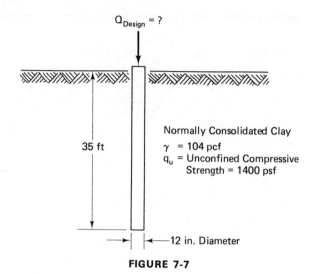

FIGURE 7-7

Solution

From Eq. (7-2),

$$Q_{\text{ultimate}} = f \cdot A_{\text{surface}} + q \cdot A_{\text{tip}} \tag{7-2}$$

$$f = \text{adhesion} = \alpha c$$

$$q_u = 1400 \text{ psf} = 0.7 \text{ ton/ft}^2$$

$$\alpha = 0.9 \quad (\text{see Fig. 7-6 with } q_u = 0.7 \text{ ton/ft}^2)$$

$$c = \frac{q_u}{2} = \frac{1400}{2} = 700 \text{ psf}$$

$$f = (0.9)(700) = 630 \text{ psf}$$

$$A_{\text{surface}} = (\pi d)(L) = (3.14)(1)(35) = 110 \text{ ft}^2$$

$$q_{\text{tip}} = cN_c \tag{7-6}$$

$$q_{\text{tip}} = (700)(9) = 6300 \text{ psf}$$

$$A_{\text{tip}} = \frac{\pi d^2}{4} = \frac{3.14}{4}(1)^2 = 0.785 \text{ ft}^2$$

$$Q_{\text{ultimate}} = (630)(110) + (6300)(0.785) = 74{,}200 \text{ lb} = 74.2 \text{ kips}$$

$$Q_{\text{design}} = \frac{74.2}{2} = 37.1 \text{ kips}$$

EXAMPLE 7-4

Given

A 12-in.-diameter concrete pile is driven at a site as shown in Fig. 7-8.

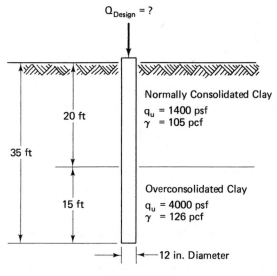

$Q_{\text{Design}} = ?$

Normally Consolidated Clay
$q_u = 1400$ psf
$\gamma = 105$ pcf

20 ft

35 ft

Overconsolidated Clay
$q_u = 4000$ psf
$\gamma = 126$ pcf

15 ft

12 in. Diameter

FIGURE 7-8

Required

The design capacity of the pile. Use a factor of safety of 2.

Solution

From Eq. (7-1),

$$Q_{\text{ultimate}} = Q_{\text{friction}} + Q_{\text{tip}} \qquad (7\text{-}1)$$

$$Q_{\text{friction}} = f \cdot A_{\text{surface}}$$

$$= f_1 \cdot A_{\text{surface}_1} + f_2 \cdot A_{\text{surface}_2}$$

With $q_{u_1} = 1400$ psf $= 0.7$ ton/ft^2 and using Fig. 7-6, $\alpha_1 = 0.9$,

$$c_1 = \frac{q_{u_1}}{2} = \frac{1400}{2} = 700 \text{ psf}$$

$$f_1 = c_1\alpha_1 = (700)(0.9) = 630 \text{ psf}$$

$$A_{\text{surface}_1} = (\pi d)(L_1) = (3.14)(1)(20) = 62.8 \text{ ft}^2$$

With $q_{u_2} = 4000$ psf $= 2.0$ tons/ft^2 and using Fig. 7-6, $\alpha_2 = 0.56$,

$$c_2 = \frac{q_{u_2}}{2} = \frac{4000}{2} = 2000 \text{ psf}$$

$$f_2 = c_2\alpha_2 = (2000)(0.56) = 1120 \text{ psf}$$

$$A_{\text{surface}_2} = (\pi d)(L_2) = (3.14)(1)(15) = 47.1 \text{ ft}^2$$

$$Q_{\text{friction}} = (630)(62.8) + (1120)(47.1) = 92,300 \text{ lb} = 92.3 \text{ kips}$$

$$N_c = 9$$

$$q_{tip} = cN_c \qquad (7\text{-}6)$$

$$q_{tip} = 2000(9) = 18{,}000 \text{ psf}$$

$$A_{tip} = \frac{\pi}{4}d^2 = \frac{3.14}{4}(1)^2 = 0.785 \text{ ft}^2$$

$$Q_{tip} = (18{,}000)(0.785) = 14{,}100 \text{ lb} = 14.1 \text{ kips}$$

$$Q_{ultimate} = 92.3 + 14.1 = 106.4 \text{ kips}$$

$$Q_{design} = \frac{106.4}{2} = 53.2 \text{ kips}$$

EXAMPLE 7-5

Given

1. A 14-in. square prestressed concrete pile is to be driven in a clay soil (see Fig. 7-9).

2. The design capacity of the pile is 80 kips.

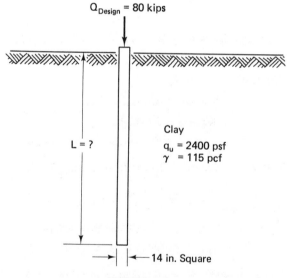

Q_{Design} = 80 kips

Clay
q_u = 2400 psf
γ = 115 pcf

L = ?

14 in. Square

FIGURE 7-9

Required

The required length of pile if the factor of safety is 2.

Solution

$$Q_{design} = 80 \text{ kips}$$

$$Q_{ultimate} = \text{F.S.} \times Q_{design} = (2)(80) = 160 \text{ kips}$$

$$c = \frac{2400}{2} = 1200 \text{ psf}$$

$$q_{tip} = cN_c = 1200(9) = 10,800 \text{ psf}$$

$$Q_{tip} = (10,800)\left(\frac{14}{12} \times \frac{14}{12}\right) = 14,700 \text{ lb} = 14.7 \text{ kips}$$

From Eq. (7-1),

$$Q_{ultimate} = Q_{friction} + Q_{tip} \tag{7-1}$$

$$Q_{friction} = Q_{ultimate} - Q_{tip}$$

$$Q_{friction} = 160 - 14.7 = 145.3 \text{ kips}$$

$$Q_{friction} = f \cdot A_{surface} = \alpha c A_{surface}$$

From Fig. 7-6, with $q_u = 2400 \text{ psf} = 1.2 \text{ tons/ft}^2$,

$$\alpha = 0.76$$

$$c = \frac{2400}{2} = 1200 \text{ psf}$$

$$145,300 = (0.76)(1200)\left(\frac{14}{12} \times 4\right)(L)$$

$$L = 34 \text{ ft}$$

The required length of the 14-in. square pile is 34 ft.

Soft clays adjacent to piles may lose a large portion of their strength as a result of being disturbed by the driving of the pile. Propitiously, the disturbed clay gains strength rapidly after driving stops. The full strength of the original clay is usually regained within a month or so after pile driving has terminated. Ordinarily, this is not a problem, since piles are not usually loaded immediately after driving, and thus the clay will have time to regain its original strength. In cases where piles are to be loaded immediately after driving, the effect of decreased strength must be taken into account by performing laboratory tests to determine the extent of the strength reduction and the rate of strength recovery [3].

7-5 PILE-DRIVING FORMULAS

In theory, it seems possible to calculate pile capacity based on the amount of energy delivered to the pile by the hammer and the resulting penetration of the pile. Intuitively, the greater the resistance required to drive a pile, the greater will be the capacity of the pile to carry load. Hence, many attempts have been made to develop "pile-driving formulas" by equating the energy delivered by the hammer to the work done by the pile as it penetrates a certain distance against a certain resistance, with allowance made for energy losses.

Generally, no pile-driving formula has been developed that gives accurate results for pile capacity. The soil resistance does not remain constant during and after the pile-driving operation. In addition, pile-driving formulas give varying results. While pile-driving formulas are not generally used to determine pile capacity, they may be used to determine when to stop driving a pile so that the bearing capacity of that pile will be the same as that of a test pile or of other piles driven in the same subsoil. These piles should be driven until the number of blows required to drive the last inch is the same as that of the test piles that furnished the information for evaluating the design load. However, piles driven in soft silt or clay should all be driven to the same depth, rather than driven a certain number of blows [1]. Penetration resistance can also be used to prevent pile damage due to overdriving.

One simple and widely used pile-driving formula is known as the Engineering-News formula. It is given by [9]

$$Q_a = \frac{2W_r H}{S + C} \tag{7-7}$$

or

$$Q_a = \frac{2E}{S + C} \tag{7-8}$$

where Q_a = allowable pile capacity, lb

W_r = weight of ram, lb

H = height of fall of ram, ft

S = amount of pile penetration per blow, in./blow

C = 1.0 for drop hammer

C = 0.1 for steam hammer

E = driving energy

The Engineering-News formula (above) has a built-in factor of safety of 6. Tests have shown that this formula is not really reliable for computing pile loads, and it should be avoided except as a rough guide [2].

Another pile-driving formula is known as the Danish formula. It is given by [2]

$$Q_{\text{ultimate}} = \frac{e_h(E_h)}{S + \frac{1}{2}S_0} \tag{7-9}$$

where Q_{ultimate} = ultimate capacity of the pile

e_h = efficiency of pile hammer (see Table 7-5)

E_h = manufacturers' hammer energy rating (see Table 7-6)

TABLE 7-5 Pile hammer efficiency [7].

Type of Hammer	Efficiency, e_h
Drop hammer	0.75–1.00
Single-acting hammer	0.75–0.85
Double-acting hammer	0.85
Diesel hammer	0.85–1.00

TABLE 7-6 Properties of selected impact pile hammers [8].

Rated Energy (ft–lb)	Make	Model	Type[1]	Blows per Minute[2]	Stroke at Rated Energy (in.)	Weight Striking Parts (lb)
7,260	Vulcan	2	S	70	29	3,000
8,750	MKT[3]	9B3	DB	145	17	1,600
13,100	MKT	10B3	DB	105	19	3,000
15,000	Vulcan	1	S	60	36	5,000
15,100	Vulcan	50C	DF	120	$15\frac{1}{2}$	5,000
16,000	MKT	DE–20	DE	48	96	2,000
18,200	Link-Belt	440	DE	86–90	$36\frac{7}{8}$	4,000
19,150	MKT	11B3	DB	95	19	5,000
19,500	Raymond	65C	DF	100–110	16	6,500
19,500	Vulcan	06	S	60	36	6,500
22,400	MKT	DE–30	DE	48	96	2,800
22,500	Delmag	D–12	DE	42–60		2,750
24,375	Vulcan	0	S	50	39	7,500
24,400	Kobe	K13	DE	45–60	102	2,870
24,450	Vulcan	80C	DF	111	16	8,000
26,000	Vulcan	08	S	50	39	8,000
26,300	Link-Belt	520	DE	80–84	$43\frac{1}{8}$	5,070
32,000	MKT	DE–40	DE	48	96	4,000
32,500	MKT	S10	S	55	39	10,000
32,500	Vulcan	010	S	50	39	10,000
32,500	Raymond	00	S	50	39	10,000
36,000	Vulcan	140C	DF	103	$15\frac{1}{2}$	14,000
39,700	Delmag	D–22	DE	42–60		4,850
40,600	Raymond	000	S	50	39	12,500
41,300	Kobe	K–22	DE	45–60	102	4,850
42,000	Vulcan	014	S	60	36	14,000
48,750	Vulcan	016	S	60	36	16,250

[1]S = single-acting steam; DB = double-acting steam; DF = differential-acting steam; DE = diesel.
[2]After development of significant driving resistance.
[3]For many years known as McKiernan–Terry.

S = average penetration of the pile from the last few driving blows

S_0 = elastic compression of the pile

$$S_0 = \left[\frac{2e_h E_h L}{AE}\right]^{1/2}$$

L = length of pile

A = cross-sectional area of pile

E = modulus of elasticity of pile material

Statistical studies indicate that a factor of safety of 3 should be used with the Danish formula.

Example 7-6 demonstrates how the Danish formula can be used as a field control during pile driving to indicate when desired pile capacity has been obtained.

EXAMPLE 7-6

Given

1. The design capacity of a 12-in. steel pipe pile is 100 kips.
2. The modulus of elasticity of the steel pipe pile is 29,000 ksi.
3. The length of the pile is 40 ft.
4. The cross-sectional area of the pipe pile is 16 in.^{2}
5. The hammer is a Vulcan 140 C with a weight of pile hammer ram of 14,000 lb and a manufacturer's hammer energy rating of 36,000 ft-lb.
6. The hammer efficiency is assumed to be 0.80.

Required

1. What should be the average penetration of the pile from the last few driving blows in in./blow?
2. How many blows/ft for the last foot of penetration are required for the design capacity, using the Danish formula?

Solution

1. From Eq. (7-9),

$$Q_{ultimate} = \frac{e_h(E_h)}{S + \frac{1}{2}S_0} \qquad (7\text{-}9)$$

$$S + \tfrac{1}{2}S_0 = \frac{e_h(E_h)}{Q_{ultimate}}$$

$$S = \frac{e_h(E_h)}{Q_{\text{ultimate}}} - \tfrac{1}{2}S_0$$

$$Q_{\text{design}} = \frac{Q_{\text{ultimate}}}{\text{F.S.}} = \frac{Q_{\text{ultimate}}}{3}$$

$$Q_{\text{ultimate}} = 3 \times Q_{\text{design}} = 3 \times 100 = 300 \text{ kips}$$

$$S_0 = \left[\frac{2e_h E_h L}{AE}\right]^{1/2}$$

$$e_h = 0.80$$

$$E_h = 36,000 \text{ ft-lb} = 36 \text{ ft-kips}$$

$$L = 40 \text{ ft}$$

$$A = 16 \text{ in.}^2$$

$$E = 29,000 \text{ ksi}$$

$$S_0 = \left[\frac{(2)(0.80)(36 \text{ ft-kips})(40 \text{ ft})}{(16 \text{ in.}^2)(29,000 \text{ kips/in.}^2)}\right]^{1/2} = 0.070 \text{ ft} = 0.84 \text{ in.}$$

$$S = \frac{(0.80)(36 \text{ ft-kips})(12 \text{ in./ft})}{300 \text{ kips}} - (\tfrac{1}{2})(0.84 \text{ in.}) = 0.73 \text{ in./blow}$$

2. Number of blows required for the last foot of penetration

$$= \frac{12 \text{ in./ft}}{0.73 \text{ in./blow}} = 16 \text{ blows/ft}$$

7-6 PILE LOAD TESTS

Load tests are performed on-site on test piles to determine or to verify the design capacity of piles. Normally, piles are designed initially by analytical or other methods, based on estimated loads and soil characteristics. The pile load tests are performed on test piles during the design stage to check the design capacity. Should load test results indicate possible bearing failure or excessive settlement, the pile design should be revised accordingly. Also, data collected from pile load tests will help develop criteria for the foundation installation.

To carry out pile load tests, the first step is to drive the test piles. They should be driven at a location where soil conditions are known (such as near a bore hole) and where soil conditions are relatively poor. Both the test piles and the method of driving them should be exactly the same as will be used in the construction project. A penetration record should be kept as the test pile is driven.

The next step is to load the test piles. For reasons explained previously in this chapter, test piles in clays should not be loaded until some time (at least several weeks) has passed after the piles were driven. Test piles in sands, however, may be loaded several days after the piles were driven. The piles may be loaded by adding dead weight or by hydraulically jacking (against a fixed platform, for example). The total test load on test piles should be 200%

of the proposed design load. The load should be applied to the pile in increments of 25% of the total test load. For specific details regarding loading,
the reader is referred to the ASTM Book of Standards. In any event a record
of load and corresponding settlement must be kept as the test pile is loaded
and unloaded.

The next step is to plot a load-settlement graph, as shown in Fig. 7-10.

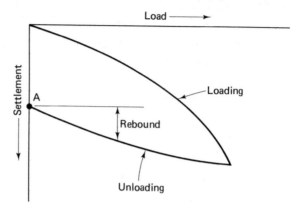

FIGURE 7-10　Typical load-settlement graph.

From this graph, the relationship between load and net settlement can be
obtained. Ordinates along the loading curve of Fig. 7-10 give gross settlement.
Subtracting the final settlement upon unloading (point A in Fig. 7-10) from
ordinates along the unloading curve gives the rebound. Net settlement can
then be determined by subtracting rebound from corresponding gross
settlement.

The allowable pile load is generally determined based on criteria specified
by the applicable building codes. There are many building codes and therefore
many criteria for determining allowable pile loads based on pile tests. It is,
of course, the responsibility of the soils engineer or technologist to follow the
criteria specified by the applicable building code. Examples 7-7 and 7-8, in
addition to illustrating the determination of allowable pile load by the pile
load test, give two possible building code criteria for determining pile capacity
by the pile load test.

EXAMPLE 7-7

Given

1. A 12-in.-diameter pipe pile with a length of 50 ft was subjected to
a pile load test.

2. The results of the test were plotted and the load-settlement curve is
shown in Fig. 7-11.

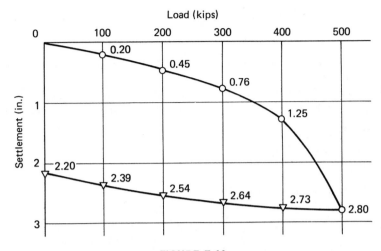

FIGURE 7-11

3. The local building code states that the allowable pile load is taken as one-half of that load which produces a net settlement of not more than 0.01 in./ton, but in no case more than 0.75 in.

Required

Allowable pile load.

Solution

Net settlement = gross settlement − rebound

Test Load (kips)	Test Load (tons)	Gross Settle- ment (in.)	Rebound (in.)	Net Settlement (in.)	Building Code Maximum Allowable Settlement (in.)
100	50	0.20	2.39 − 2.20 = 0.19	0.20 − 0.19 = 0.01	0.5
200	100	0.45	2.54 − 2.20 = 0.34	0.45 − 0.34 = 0.11	~~1.0~~(use 0.75)
300	150	0.76	2.64 − 2.20 = 0.44	0.76 − 0.44 = 0.32	~~1.5~~(use 0.75)
400	200	1.25	2.73 − 2.20 = 0.53	1.25 − 0.53 = 0.72	~~2.0~~(use 0.75)
500	250	2.80	2.80 − 2.20 = 0.60	2.80 − 0.60 = 2.20	~~2.5~~(use 0.75)

Since a test load of 200 tons produces a net settlement of 0.72 in. and the maximum allowable settlement is 0.75 in.,

$$\text{Allowable load} = \frac{200}{2} = 100 \text{ tons}$$

EXAMPLE   7-8

Given

The same conditions as in Example 7-7 except that another local building code is to be applied as follows: "The allowable pile load shall be not more than one-half of that test load which produces a net settlement per ton of test load of not more than 0.01 in. but in no case more than one-half inch."

Required

Allowable pile load.

Solution

From previous example:

Test Load (*tons*)	Net Settlement (*in.*)	Building Code Maximum Allowable Settlement (*in.*)
50	0.01	0.5
100	0.11	~~1.0~~ (use 0.5)
150	0.32	~~1.5~~ (use 0.5)
200	0.72	~~2.0~~ (use 0.5)
250	2.20	~~2.5~~ (use 0.5)

Since a test load of 150 tons produces a net settlement of 0.32 in. and the maximum allowable settlement is 0.5 in.,

$$\text{Allowable load} = \frac{150}{2} = 75 \text{ tons}$$

7-7   NEGATIVE SKIN FRICTION (DOWNDRAG)

As related throughout this chapter, piles depend, in part at least, on skin friction for support. Under certain conditions, however, skin friction may develop that causes downdrag on the pile rather than support. Skin friction that causes downdrag is known as *negative skin friction.*

Negative skin friction may occur if the soil adjacent to a pile settles more than the pile itself. This is most likely to happen when the pile is driven through compressible soil, such as soft to medium clays or soft silt. Subsequent consolidation of the soil (caused by newly placed fill, for example) can cause negative skin friction as the soil adjacent to a pile moves downward while the pile, restrained at the tip, remains fixed. A similar phenomenon may occur as a result of a lowering of the water table at the site.

Negative skin friction is, of course, detrimental with regard to a pile's ability to carry load. Hence, if conditions at a particular site suggest that negative skin friction may occur, its magnitude should be determined and subtracted from the load-carrying ability of the pile.

7-8 PILE GROUPS AND SPACING OF PILES

Heretofore in this chapter, discussion has pertained to a single pile. In reality, however, piles are almost always arranged in groups of three or more. Furthermore, the group of piles is commonly tied together by a pile cap, which is attached to the heads of the individual piles and causes the several piles to act together as a pile foundation. Figure 7-12 illustrates some typical pile grouping patterns.

If two piles are driven close together, the soil stresses caused by the piles tend to overlap; and the bearing capacity of the two pile group is less than the sum of the individual capacities. If the piles are moved farther apart, so that the individual stresses do not overlap, the bearing capacity of the two pile group is not reduced significantly from the sum of the individual capacities. Thus, it would appear that piles should be spaced relatively far apart. This consideration is offset, however, by the unduly large pile caps that would be required (for the wider spacing).

The minimum allowable pile spacing is often specified by applicable building codes. For example, a building code may state that "the minimum center-to-center spacing of piles not driven to rock shall be not less than twice the average diameter of a round pile, nor less than 1.75 times the diagonal dimension of a rectangular or rolled structural steel pile, nor less than 2 ft 6 in. For piles driven to rock, the minimum center-to-center spacing of piles shall be not less than twice the average diameter of a round pile, nor less than 1.75 times the diagonal dimension of a rectangular or rolled structural steel pile, nor less than 2 ft 0 in." [10].

7-9 EFFICIENCY OF PILE GROUPS

As related in the last section, the capacity of a group of piles may be less than the sum of the individual capacities of the piles making up the group. Inasmuch as it would be convenient to estimate the capacity of a group of piles based on the capacity of a single pile, attempts have been made to be able to determine the efficiency of pile groups. (Efficiency of a pile group is the capacity of a group of piles divided by the sum of the individual capacities of the piles making up the group.)

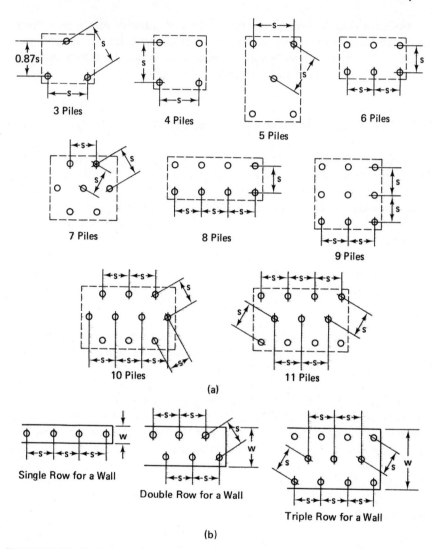

FIGURE 7-12 Typical pile group patterns. (a) for single footings; (b) for foundation walls. [7]

In the case where a group of piles is comprised of end-bearing piles resting on bedrock (or on a layer of dense sand and gravel overlying bedrock), an efficiency of 1.0 may be assumed [11]. (In other words, the group of n piles will carry n times the capacity of a single pile.) An efficiency of 1.0 is also often assumed by designers for friction piles driven in cohesionless soil. In the case where a group of piles is comprised of friction piles driven in cohesive soil, an efficiency of less than 1.0 is to be expected because stresses from individual piles build up and reduce the capacity of the pile group.

One equation that has been used to compute pile group efficiency is known as the Converse–Labarre equation [11]:

$$E_g = 1 - \theta \frac{(n-1)m + (m-1)n}{90mn} \tag{7-10}$$

where E_g = pile group efficiency

         θ = arctan d/s, deg

         n = number of piles in a row

        m = number of rows of piles

        d = diameter of piles

        s = spacing of piles, center to center, in same units as pile diameter

Example 7-9 illustrates the application of the Converse–Labarre equation.

EXAMPLE 7-9

Given

1. A pile group consists of 12 friction piles in cohesive soil (see Fig. 7-13).
2. The diameter of each pile is 12 in. and center-to-center spacing is 3 ft.

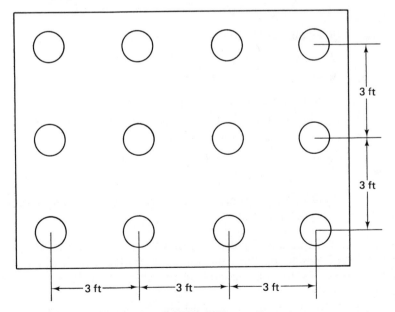

FIGURE 7-13

3. By means of a load test, the ultimate load of a single pile was found to be 100 kips.

Required

Design capacity of the pile group, using the Converse–Labarre equation.

Solution

$$E_g = 1 - \theta \frac{(n-1)m + (m-1)n}{90mn} \qquad (7\text{-}10)$$

$$\theta = \arctan \frac{d}{s} = \arctan \frac{1}{3} = 18.4°$$

$$E_g = 1 - (18.4)\frac{(4-1)(3) + (3-1)(4)}{(90)(3)(4)} = 0.71$$

Allowable bearing value of a single pile $= \dfrac{100}{2} = 50$ kips

Design capacity of the pile group $= (0.71)(12)(50) = 426$ kips

For friction piles driven in cohesive soil, pile group efficiency may be assumed to vary linearly from a value of 0.7 at a pile spacing of 3 times the pile diameter to a value of 1.0 at a pile spacing of 8 times the pile diameter [12, 13]. For pile spacings less than 3 times the pile diameter, the group capacity may be considered as block capacity and the total capacity can be estimated by treating the group as a pier and applying the equation [1, 12]

$$Q_g = 2D(W + L)f + 1.3 \times c \times N_c \times W \times L \qquad (7\text{-}11)$$

where $Q_g =$ ultimate bearing capacity of the pile group

$D =$ depth of the pile group

$W =$ width of the pile group

$L =$ length of the pile group

$f =$ unit adhesion or skin friction developed between cohesive soil and pile surface (equal to αc)

$\alpha =$ ratio of adhesion to cohesion (see Fig. 7-6)

$c =$ cohesion

$N_c =$ bearing capacity coefficient for a shallow rectangular footing (see Fig. 6-6)

A pile group can be considered safe against block failure if total design load (i.e., the "safe design load" per pile multiplied by the number of piles) does not exceed $Q_g/3$. If the total design load exceeds $Q_g/3$, the foundation design must be revised.

Figure 7-14 gives a summary of criteria for pile group capacity.

Individual Pile Failure   Block Failure

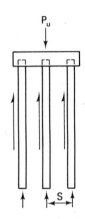

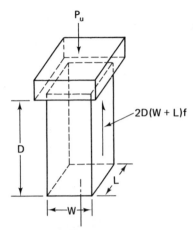

In Cohesionless Soils

$$P_u = n \times Q_u$$

In Cohesive Soils

for $S \geq 3$ Diameters,

$P_u = E \times n \times Q_u$

E Varies Linearly from
0.7 at $S = 3$, to 1.0 at $S \geq 8$

In Cohesive Soils

for $S < 3.0$,

$P_u = 2D(W + L)f + 1.3 \times c \times N_c \times W \times L$

FIGURE 7-14 Summary of criteria for pile group capacity. [12, 13]

EXAMPLE 7-10

Given

    1.   A pile group consists of 4 friction piles in cohesive soil (see Fig. 7-15).

    2.   The diameter of each pile is 12 in. and center-to-center spacing is 2.5 ft.

Required

    1.   The block capacity of the pile group. Use a factor of safety of 3.

    2.   The allowable group capacity based on individual pile failure. Use a factor of safety of 2, along with the Converse–Labarre equation for pile group efficiency.

    3.   Design capacity of the pile group.

Solution

    1.   *Block capacity:*
Since the center-to-center spacing of the piles is 2.5 ft, which is less than 3 ft (i.e., 3 diameters), according to the criteria suggested by

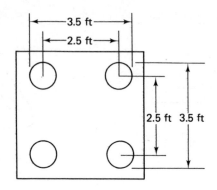

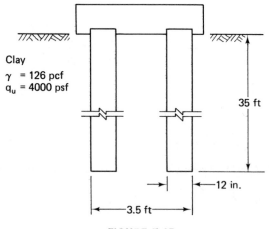

FIGURE 7-15

Coyle and Sulaiman, the block capacity of the pile group can be estimated by Eq. (7-11).

$$Q_g = 2D(W + L)f + 1.3 \times c \times N_c \times W \times L \qquad (7\text{-}11)$$

$$D = 35 \text{ ft}$$

$$W = 2.5 + 0.5 + 0.5 = 3.5 \text{ ft}$$

$$L = 2.5 + 0.5 + 0.5 = 3.5 \text{ ft}$$

$$f = \alpha c$$

$$q_u = 4000 \text{ psf} = 2.0 \text{ tons/ft}^2$$

$$c = \frac{4000}{2} = 2000 \text{ psf} = 2 \text{ ksf}$$

From Fig. 7-6 with $q_u = 2.0$ tons/ft²,

$\alpha = 0.56$

$f = (0.56)(2000) = 1120$ psf $= 1.12$ ksf

$N_c = 5.14$ (from Fig. 6-6 with $\phi = 0°$ for clay)

$Q_g = (2)(35)(3.5 + 3.5)(1.12) + (1.3)(2)(5.14)(3.5)(3.5)$
$= 713$ kips

$$\text{Allowable block capacity} = \frac{713}{3} = 238 \text{ kips}$$

2. *Group capacity based on individual pile:*

$Q_{\text{ultimate}} = Q_{\text{friction}} + Q_{\text{tip}}$ \hfill (7-1)

$Q_{\text{friction}} = f \cdot A_{\text{surface}}$

$f = 1.12$ ksf [from (1) above]

$A_{\text{surface}} = (\pi d)(L) = (3.14)(1)(35) = 109.9$ ft²

$Q_{\text{friction}} = (1.12)(109.9) = 123$ kips

$$Q_{\text{tip}} = cN_cA_{\text{tip}} = (2)(9)\left(\frac{\pi}{4}\right)(1)^2 = 14 \text{ kips}$$

$Q_{\text{ultimate}} = 123 + 14 = 137$ kips

$$Q_a = \frac{137}{2} = 68.5 \text{ kips}   \text{(allowable load for an individual pile)}$$

$$E_g = 1 - \theta\frac{(n-1)m + (m-1)n}{90mn} \hfill (7\text{-}10)$$

$$\theta = \arctan\frac{d}{s} = \arctan\frac{1}{2.5} = 21.8°$$

$n = 2$

$m = 2$

$$E_g = 1 - (21.8)\frac{(2-1)(2) + (2-1)(2)}{(90)(2)(2)} = 0.758$$

Allowable $Q = (68.5)(4)(0.758) = 208$ kips   (allowable load for pile group)

3. *Design capacity of the pile group:*
This will be the smaller group capacity of (1) and (2), which is 208 kips.

7-10 DISTRIBUTION OF LOADS IN PILE GROUPS

The load on any particular pile within a pile group may be computed using the elastic equation [3]:

$$Q_m = \frac{Q}{n} \pm \frac{M_y x}{\sum (x^2)} \pm \frac{M_x y}{\sum (y^2)} \hfill (7\text{-}12)$$

where Q_m = axial load on any pile m

Q = total vertical load acting at the centroid of the pile group

n = number of piles

M_x, M_y = moment with respect to x and y axes, respectively

x, y = distance from pile to y and x axes, respectively

(Both x and y axes pass through the centroid of the pile group and are per-pendicular to each other.) It should be noted that shears and bending moments can be determined for any section of pile cap by using elastic and static equations.

EXAMPLE 7-11

Given

1. A pile group consists of 9 piles as shown in Fig. 7-16.
2. A column load of 450 kips acts vertically on point A.

Required

Determine the loads on pile Nos. 1, 6, and 8.

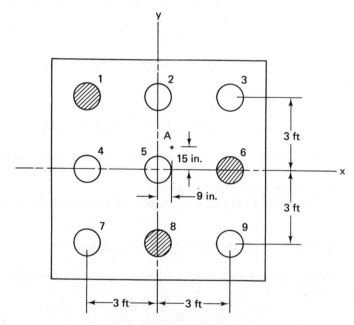

FIGURE 7-16

Solution

From Eq. (7-12),

$$Q_m = \frac{Q}{n} \pm \frac{M_y x}{\sum (x^2)} \pm \frac{M_x y}{\sum (y^2)} \tag{7-12}$$

$$Q = 450 \text{ kips}$$

$$n = 9$$

$$\sum x^2 = (6)(3)^2 = 54 \text{ ft}^2$$

$$\sum y^2 = (6)(3)^2 = 54 \text{ ft}^2$$

$$M_x = (450)\left(\frac{15}{12}\right) = 562.5 \text{ kip-ft}$$

$$M_y = (450)\left(\frac{9}{12}\right) = 337.5 \text{ kip-ft}$$

Load on pile No. 1

$$Q_1 = \frac{450}{9} + \frac{(337.5)(-3)}{54} + \frac{(562.5)(+3)}{54} = 62.5 \text{ kips}$$

Load on pile No. 6

$$Q_6 = \frac{450}{9} + \frac{(337.5)(+3)}{54} + \frac{(562.5)(0)}{54} = 68.8 \text{ kips}$$

Load on pile No. 8

$$Q_8 = \frac{450}{9} + \frac{(337.5)(0)}{54} + \frac{(562.5)(-3)}{54} = 18.8 \text{ kips}$$

EXAMPLE 7-12

Given

1. Figure 7-17 shows a pile foundation consisting of 5 piles.
2. The pile foundation is subjected to a 200-kip vertical load and a moment with respect to the y axis of 140 kip-ft (see Fig. 7-17).

Required

Determine the shear and bending moment on section a–a due to the pile reacting under the pile cap.

Solution

From Eq. (7-12),

$$Q_m = \frac{Q}{n} \pm \frac{M_y x}{\sum (x^2)} \pm \frac{M_x y}{\sum (y^2)} \tag{7-12}$$

$$Q = 200 \text{ kips}$$

$$n = 5$$

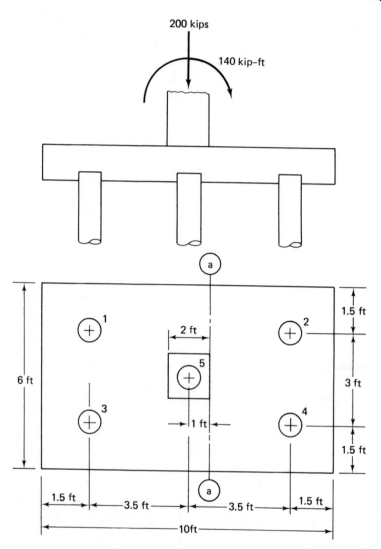

FIGURE 7-17

$$M_y = 140 \text{ kip-ft}$$

$$M_x = 0$$

$$\sum (x^2) = (4)(3.5)^2 = 49$$

$$Q_2 = Q_4 = \frac{200}{5} + \frac{(140)(3.5)}{49} + \frac{(0)y}{\sum y^2} = 50 \text{ kips}$$

Shear at section a–a = (50)(2) = 100 kips

Moment at section a–a = (2)(50)(3.5 − 1) = 250 kip-ft

7-11 SETTLEMENT OF PILE FOUNDATIONS

Like shallow foundations, pile foundations must be analyzed to predict their settlement to ensure that it is tolerable. Unfortunately, universally accepted methods for predicting pile settlement are not available today. The following give some possible methods for predicting pile settlement for end–bearing piles on bedrock, piles in sand, and piles in clay.

Settlement of End-bearing Piles on Bedrock

A well-designed and constructed pile foundation on hard bedrock generally will not experience an objectionable amount of settlement. The amount of settlement of pile foundations on soft bedrock is very difficult to predict accurately and can only be estimated by judging from the characteristics of the rock core samples. Local experience, if available, should be employed as guidance [3].

Settlement of Piles in Sand

Settlement of a pile group in sand may be estimated by using the empirical relationship of Fig. 7-18. This relationship gives the ratio of settlement of pile group to settlement of single test pile as a function of the width of the pile group in sand.

The settlement of a pile group is substantially larger than that of a single test pile, which can readily be determined by a pile load test. For example,

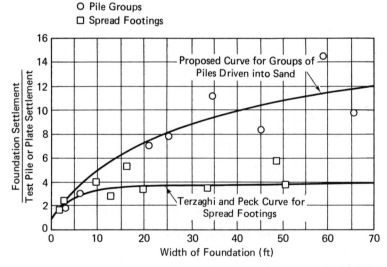

FIGURE 7-18 Settlement ratio versus width of foundations on sand. [12, 14]

according to Fig. 7-18 settlement of a pile group with a 15-ft width would have about 6.4 times that of a single test pile.

Settlement of Piles in Clay

Prediction of settlement of piles in deep clay requires first an estimate of load distribution in the soil, followed by settlement calculation in accordance with consolidation theory. One method of estimating the load distribution is to assume the load is applied to an equivalent flexible mat (i.e., an imaginary mat) at some selected level and then to compute the distribution of load from the imaginary mat. For friction piles in deep clay, the equivalent (imaginary) mat may be assumed at a plane located at two-thirds the pile depth [1, 12] (see Fig. 7-19a). Consolidation of the soil below that plane is then computed as if the piles are no longer present. If the piles pass through a layer of very soft clay to a firm bearing in a layer of stiff clay, an equivalent mat may be placed at the level of the pile tips, assuming eventual concentration of load at that level (see Fig. 7-19b) [12].

Settlement analysis is then performed, based on consolidation test results to predict the expected, approximate settlement that would occur for an ordinary (unpiled) foundation if the foundation were a mat of the same depth and dimensions at the same plane. In such cases the method of settlement analysis of pile-supported foundations is the same as that used for shallow foundations. From Chap. 4, based on consolidation test results, the amount of settlement due to consolidation can be calculated for a layer of compressible soil by the equation [1, 3]

$$S = \frac{e_0 - e}{1 + e_0}[H] \qquad (4\text{-}3)$$

or

$$S = C_c \frac{H}{1 + e_0} \log \frac{P_0 + \Delta p}{P_0} \qquad (4\text{-}4)$$

where S = consolidation settlement

e_0 = initial void ratio (void ratio *in situ*)

e = final void ratio

H = thickness of layer of compressible soil

C_c = compression index (slope of field **e–log p** curve)

P_0 = effective overburden pressure (effective weight of soil above midheight of the consolidating layer)

Δp = consolidation pressure (net additional pressure)

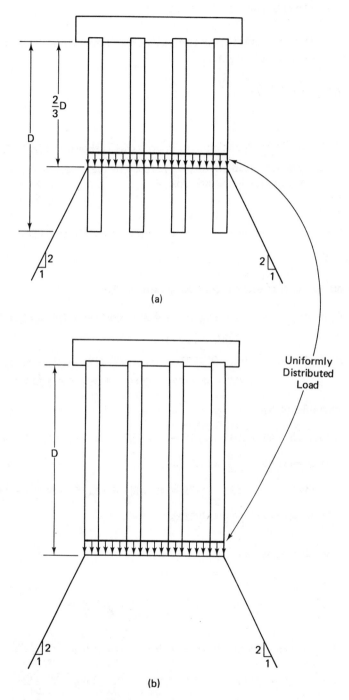

FIGURE 7-19 (a) Friction piles in deep clay. (b) Friction piles through soft clay into stiff clay. [12]

Example 7-13 illustrates the computation of approximate total settlement of a pile foundation in deep clay.

EXAMPLE 7-13

Given

1. A group of friction piles in deep clay is shown in Fig. 7-20.
2. The total load on the piles reduced by the weight of soil displaced by the foundation is 300 kips.

Required

Calculate the approximate total settlement of the pile foundation.

Solution

Computation of effective overburden pressures (P_0)

P_0 at elev. 66 $= (100 - 95)(102) + (95 - 89)(119) + (89 - 66)(119 - 62.4)$

$\qquad = 2526$ psf $= 2.53$ ksf

P_0 at elev. 49 $= (100 - 95)(102) + (95 - 89)(119) + (89 - 57)(119 - 62.4)$

$\qquad + (57 - 49)(125 - 62.4) = 3536$ psf $= 3.54$ ksf

Computation of Δp

Area at elev. 66 $= [10 + (2)(75 - 66)(\frac{1}{2})][7 + (2)(75 - 66)(\frac{1}{2})] = 304$ ft^2

Δp at elev. 66 $= \dfrac{300}{304} = 0.99$ ksf

Area at elev. 49 $= [10 + (2)(75 - 49)(\frac{1}{2})][7 + (2)(75 - 49)(\frac{1}{2})] = 1188$ ft^2

Δp at elev. 49 $= \dfrac{300}{1188} = 0.25$ ksf

Settlement computations

From Eq. (4-4),

$$S = C_c \frac{H}{1 + e_0} \log \frac{P_0 + \Delta p}{P_0} \qquad (4\text{-}4)$$

Elev. 75 to 57: $\quad S = (0.24)\left(\dfrac{18}{1 + 0.78}\right) \log \dfrac{2.53 + 0.99}{2.53} = 0.35$ ft

Elev. 57 to 41: $\quad S = (0.20)\left(\dfrac{16}{1 + 0.67}\right) \log \dfrac{3.54 + 0.25}{3.54} = 0.06$ ft

Approximate total settlement $= 0.35 + 0.06 = 0.41$ ft $= 4.9$ in.

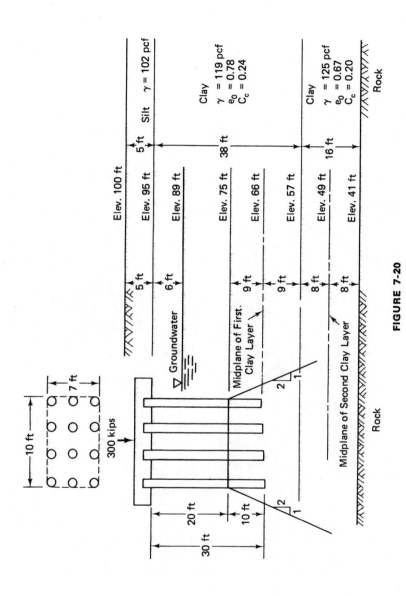

FIGURE 7-20

TABLE 7-7 Data for selection of pile hammers for driving concrete, timber, and steel sheetpiling under average and heavy driving conditions[1] [3].

Length of Pile (ft)	Depth of Penetration (%)	SHEET PILE[2] (ft-lb per blow)			TIMBER PILE (ft-lb per blow)		CONCRETE PILE (ft-lb per blow)	
		Light	Medium	Heavy	Light	Heavy	Light	Heavy
Driving through earth, sand, loose gravel—normal frictional resistance								
25	50	1000–1800	1000–1800	1800–2500	3600–4200	3600–7250	7250–8750	8750–15,000
	100	1000–3600	1800–3600	1800–3600	3600–7250	3600–8750	7250–8750	13,000–15,000
50	50	1800–3600	1800–3600	3600–4200	3600–8750	7250–8750	8750–15,000	13,000–25,000
	100	3600–4200	3600–4200	3600–7500	7250–8750	7250–15,000	13,000–15,000	15,000–25,000
75	50		3600–7500	3600–8750		13,000–15,000		19,000–36,000
	100			3600–8750		15,000–19,000		19,000–36,000
Driving through stiff clay, compacted gravel—very resistant								
25	50	1800–2500	1800–2500	1800–4200	7250–8750	7250–8750	7250–8750	8750–15,000
	100	1800–3600	1800–3600	1800–4200	7250–8750	7250–8750	7250–15,000	13,000–15,000
50	50	1800–4200	3600–4200	3600–8750	7250–15,000	7250–15,000	13,000–15,000	13,000–25,000
	100		3600–8750	3600–13,000		13,000–15,000		19,000–36,000
75	50		3600–8750	3600–13,000		13,000–15,000		19,000–36,000
	100			7500–19,000		15,000–25,000		19,000–36,000
Weight (per lin. ft)		20 lb	30 lb	40 lb	30 lb	60 lb	150 lb	400 lb
Pile size (approx.)		15 in.	15 in.	15 in.	13 in. diam	18 in. diam	12 in.²	20 in.²

[1]Tennessee Valley Authority.
[2]Energy required in driving single-sheet pile. Double these when driving two piles at a time.

7-12 CONSTRUCTION OF PILE FOUNDATIONS

Construction of pile foundations consists of installing the piles (usually by driving) and constructing the pile caps. Pile caps are often made of concrete, and their construction is usually a relatively simple structural problem.

With regard to the installation of the piles, most piles are driven by a device called a *pile hammer*. Simply speaking, a pile hammer is a weight that is alternately raised and dropped onto the top of the pile to drive the pile into the soil. Hammer weights vary considerably. As a general rule, the weight of the hammer should be at least one-half the weight of the pile being driven, and the driving energy should be at least 1 ft-lb for each pound of pile weight [3]. The hammer itself is contained within a larger device, with the hammer operated between a pair of parallel steel members known as leads.

Several types of pile hammers are available. *Drop hammers* consist of a heavy ram that is raised by a cable and hoisting drum and dropped onto the pile. *Single-acting hammers* consist of a heavy ram that is raised by steam or compressed air and dropped onto the pile. *Double-acting hammers* consist of a heavy ram that is both raised and accelerated downward by steam or air. *Differential-acting hammers* are similar to double-acting hammers. *Diesel hammers* use gasoline for fuel, which causes an explosion that advances the pile and lifts the ram. The total driving energy delivered to the pile includes both the impact of the ram and the energy delivered by explosion. Table 7-6 (in Sec. 7-5) gives more specific information on various pile hammers.

The selection of a pile hammer for a specific job depends on a number of factors. Table 7-7 gives data for selection of pile hammers for various conditions.

7-13 PROBLEMS

7-1 A 12-in. square concrete pile is driven into a loose sand to a depth of 30 ft. The soil conditions are shown in Fig. 7-21. What is the approximate axial capacity of the pile if K is assumed to be 0.7 and the factor of safety is 2?

7-2 Rework Problem 7-1 assuming the groundwater table is located 5 ft below the ground surface

7-3 A 14-in. square concrete pile is driven at a site as shown in Fig. 7-22. The embedded length of pile is 40 ft. Estimate the design capacity of the pile, using a factor of safety of 2.

7-4 A 12-in.-diameter concrete pile is driven at a site as shown in Fig. 7-23. Estimate the design capacity of the pile if the factor of safety is 2.

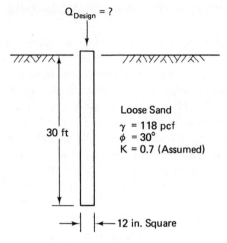

Q_{Design} = ?

30 ft

Loose Sand
γ = 118 pcf
ϕ = 30°
K = 0.7 (Assumed)

12 in. Square

FIGURE 7-21

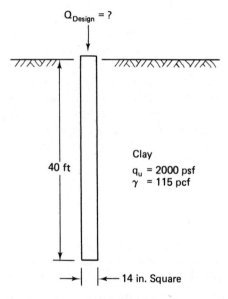

Q_{Design} = ?

40 ft

Clay
q_u = 2000 psf
γ = 115 pcf

14 in. Square

FIGURE 7-22

7-5 A 12-in.-diameter concrete pile is to be driven into a clay soil (see Fig. 7-24). The design capacity of the pile is 30 tons. Determine the required length of the pile if the factor of safety is 2.

7-6 A steel pipe pile is to be driven to an allowable load (design load) of 35 tons capacity by an MKT-11B3 double-acting steam hammer. The steel pipe has a net

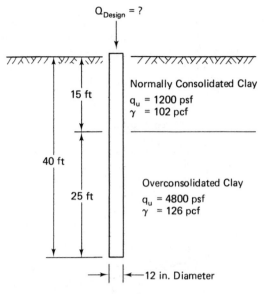

Q_{Design} = ?

15 ft

40 ft

25 ft

Normally Consolidated Clay
q_u = 1200 psf
γ = 102 pcf

Overconsolidated Clay
q_u = 4800 psf
γ = 126 pcf

12 in. Diameter

FIGURE 7-23

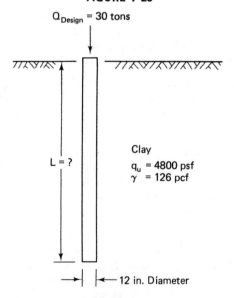

Q_{Design} = 30 tons

L = ?

Clay
q_u = 4800 psf
γ = 126 pcf

12 in. Diameter

FIGURE 7-24

cross-sectional area of 17.12 in.[2] and a length of 45 ft. The Danish pile-driving formula is to be used to control field installation of the piles. What is the required number of blows per foot for the last foot of penetration?

7-7 Rework Problem 7-6 using the Engineering-News formula.

7-8   A pile load test produces the settlement and rebound curves given in Fig. 7-25. The pile is a 12-in.-diameter concrete pile that is 25 ft long. Determine the allowable load for this pile by using a local building code which states that: "The allowable load shall not be more than one-half of that test load which produces a net settlement per ton of test load of not more than 0.01 in., but in no case more than 0.75 in."

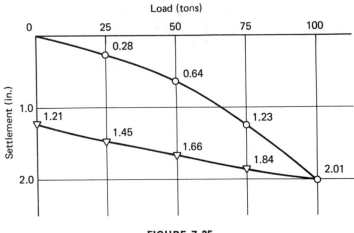

FIGURE 7-25

7-9   Rework Problem 7-8, except that the local building code is changed to read as follows: "The allowable pile load is taken as one-half of that load which produces a net settlement of not more than 0.01 in./ton of test load, but in no case more than 0.5 in."

7-10   A pile group consists of nine friction piles in clay soil. The diameter of each pile is 16 in. and the embedded length of each pile is 30 ft. Center-to-center pile spacing is 4 ft. The soil conditions are shown in Fig. 7-26. Estimate the design capacity of the pile group if the factor of safety is 2. Use the Converse–Labarre equation for pile efficiency calculation.

7-11   A 9-pile group consists of 12-in.-diameter friction concrete piles 30 ft long. The piles are driven into clay, the unconfined compressive strength of which is 6000 psf and the unit weight of which is 125 pcf. The pile spacing is $2\frac{1}{2}$ diameters (i.e., pile spacing is 2.5 ft). Find: (1) the block capacity of the pile group using a factor of safety of 3; (2) the allowable group capacity based on individual pile failure using a factor of safety of 2 along with the Converse–Labarre equation for pile group efficiency; (3) the design capacity of the pile group.

7-12   A pile group consists of 12 piles as shown in Fig. 7-27. A vertical load of 480 kips acts vertically on point *A*. Determine the load on piles Nos. 2, 4, 7, and 9.

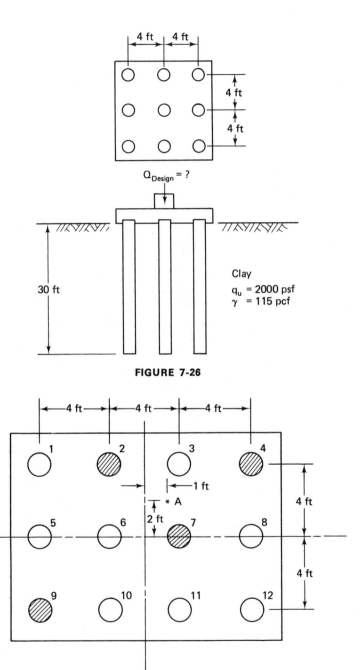

FIGURE 7-26

FIGURE 7-27

7-13   A tower shown in Fig. 7-28 is subjected to a wind pressure of 25 psf on the projected area. The tower and its foundation weigh 320 kips. Determine the maximum and minimum pile reactions for the layout shown.

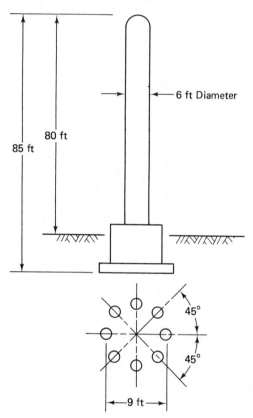

8 Piles Equally Spaced on a Circle of 9-ft Diameter

FIGURE 7-28

7-14 A group of friction piles in deep clay is shown in Fig. 7-29. The total load on the piles reduced by the weight of soil displaced by the foundation is 400 kips. Estimate the approximate total settlement of the pile foundation.

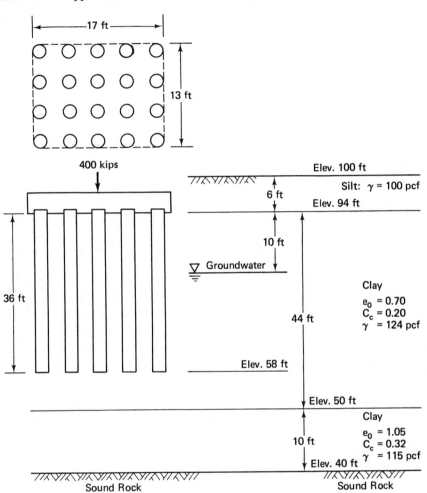

FIGURE 7-29

References

[1] KARL TERZAGHI AND RALPH B. PECK, *Soil Mechanics in Engineering Practice*, John Wiley & Sons, Inc., New York, 1967. Copyright © 1967, by John Wiley & Sons, Inc. Reprinted by permission of John Wiley & Sons, Inc.

[2] DAVID F. MCCARTHY, *Essentials of Soil Mechanics and Foundations*, Reston Publishing Company, Inc., Reston, Va., 1977.

[3] WAYNE C. TENG, *Foundation Design*, Prentice-Hall, Inc., Englewood Cliffs, N.J., 1962.

[4] G. G. MEYERHOF, "Bearing Capacity and Settlement of Pile Foundations," *J. Geotech. Eng. Div.*, *ASCE*, **102** (GT3), (1976).

[5] A. S. VESIC, "Ultimate Loads and Settlement of Deep Foundations in Sand," *Proc. Bearing Capacity Settlement Found. Symp.*, Duke University, Durham, N.C., 1967.

[6] A. S. VESIC, "Test on Instrumental Piles, Ogeechee River Site," *J. Soil Mech. Found. Div.*, *ASCE*, **96** (SM2), (1970).

[7] JOSEPH E. BOWLES, *Foundation Analysis and Design*, 2nd ed., McGraw-Hill Book Company, New York, 1977.

[8] RALPH B. PECK, WALTER E. HANSEN, AND THOMAS H. THORNBURN, *Foundation Engineering*, 2nd ed., John Wiley & Sons, Inc., New York, 1974. Copyright © 1974, by John Wiley & Sons, Inc. Reprinted by permission of John Wiley & Sons, Inc.

[9] R. H. KAROL, *Soils and Soil Engineering*, Prentice-Hall, Inc., Englewood Cliffs, N.J., 1960.

[10] *North Carolina State Building Code*, Vol. I, *General Construction*, 1967 ed.

[11] ALFREDS R. JUMIKIS, *Foundation Engineering*, Intext Educational Publishers, Scranton, Pa., 1971. Copyright © 1971 by Harper & Row, Publishers, Inc. Reprinted by permission of the publisher.

[12] BRAMLETT MCCLELLAND, "Design and Performance of Deep Foundations," *Proc. Specialty Conf. Perform. Earth Earth-supported Struct.*, *ASCE*, **2** (June 1972).

[13] HARRY M. COYLE AND IBRAHIM H. SULAIMAN, *Bearing Capacity of Foundation Piles: State of the Art*, Highway Res. Board, Record No. 333, 1970.

[14] ALEX W. SKEMPTON, Discussion, Session 5, "Piles and Pile Foundations, Settlements of Pile Foundations," *Proc., Third Int. Conf. Soil Mech. Found. Eng., Switzerland*, **3**, (1953).

8
Drilled Caissons

8-1 INTRODUCTION

A drilled caisson is a type of deep foundation that is constructed in place by drilling a hole into the soil, often to bedrock or a hard stratum, and subsequently placing concrete in the hole. The concrete may or may not contain reinforcing steel. Some drilled caissons have straight sides throughout (straight-shaft caisson); others are constructed with enlarged bases (belled caisson) (see Fig. 8-1). The enlarged base area results in a decreased contact pressure (soil pressure) at the base of the caisson.

The purpose of a drilled caisson is to transmit a structural load to the base of the caisson, which may be bedrock or other hard stratum. In essence, a drilled caisson is primarily a compression member with an axial load applied at the top, a reaction at the bottom, and lateral support along the sides.

Drilled caissons are constructed by using auger drill equipment to form the hole in the soil. The soil is removed from the hole during drilling in contrast to the driven pile, which only compresses the soil aside. Thus, such problems as shifting and lifting of driven piles do not occur with drilled caissons. In some cases, such as in dry, strong cohesive soil, the hole may be drilled dry and without any side support. In this case, the concrete placed in the hole directly makes contact with the soil forming the sides of the hole. If cohesionless soil and/or groundwater is encountered, a bentonite slurry may be introduced during drilling to prevent the soil from caving in. (Protective casing may also be used to prevent cave in.) In this case the concrete is placed from the bottom up so as to displace the slurry. If a casing is used, it would be slowly removed as the concrete is placed. In this case, the operator makes

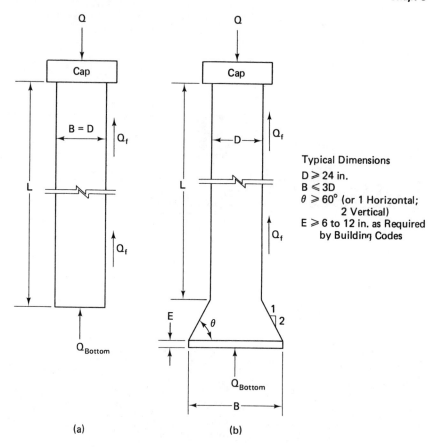

FIGURE 8-1 Caissons: (a) straight-shaft; (b) belled.

sure that the soil does not fall into the excavated hole and mix with the concrete.

Drilled caissons are a popular type of deep foundation for several reasons. The drilling equipment is relatively light and easy to use compared to pile-driving equipment. Drilling equipment is much quieter than pile drivers and does not cause the massive ground vibrations that can adversely affect adjacent piles. Finally, the drilled holes afford better (visual) inspection of the subsoil encountered.

8-2 BEARING CAPACITY OF DRILLED CAISSONS

As with a pile, a caisson gets its supporting power from two sources—skin friction and bearing capacity at the base of the caisson. Thus, at failure the load on a drilled caisson (as for a pile) may be expressed by Eq. (7-1), which is reproduced here.

$$Q_{\text{ultimate}} = Q_{\text{friction}} + Q_{\text{tip}} \qquad (7\text{-}1)$$

Since caissons are not driven (as are piles), they do not make tight contact with the surrounding soil (as do piles). Consequently, the supporting power for a caisson provided by skin friction is relatively small.

To evaluate the bearing capacity of drilled caissons, it is helpful to consider separately: drilled caissons in cohesive soils, drilled caissons in sand, and drilled caissons on rock.

Drilled Caissons in Cohesive Soils

The analysis of drilled caissons in cohesive soils is similar to that of piles in that the caisson's total bearing capacity is due to the resistance provided by its end bearing and skin friction, in accordance with Eq. (7-1). The term Q_{tip} of Eq. (7-1) can be evaluated by multiplying the cohesion of the soil (c) at the bottom of the caisson by the bearing capacity factor (N_c) and this by the area of the bottom of the caisson. The term Q_{friction} of Eq. (7-1) can be evaluated by multiplying the unit adhesion or skin friction developed between the shaft surface and the soil (f) by the surface area of the shaft (A_{shaft}) (obtained by multiplying the circumference of the caisson shaft by the depth of caisson from ground surface to the top of the bell). Making these substitutions in Eq. (7-1) gives [1]

$$Q_{\text{total}} = cN_cA_{\text{bottom}} + fA_{\text{shaft}} \qquad (8\text{-}1)$$

Thus far in this discussion, the analysis has been approximately the same as that for piles driven in clay. There is a significant difference between the two, however, and that difference is in the determination of the bearing capacity factor (N_c) and the unit adhesion or skin friction (f) of Eq. (8-1). In the case of drilled caissons, the value of the bearing capacity factor (N_c) can be obtained from Table 8-1, and the value of the unit adhesion or skin

TABLE 8-1 Bearing capacity factors [2].

Ratio of Depth of Caisson to Diameter of Caisson Bottom	N_c
0	6.2
0.5	7.1
1.0	7.7
1.5	8.1
2.0	8.4
2.5	8.6
3.0	8.8
4.0 and over	9.0

TABLE 8-2 Adhesion or skin friction values for drilled caisson foundations in clay [1].

Foundation Type and Drilling Method Utilized	Adhesion or Skin Friction, f	Upper Limit on f value (ksf)
Straight shaft, excavation drilled dry	$0.5c^1$	1.8
Straight shaft, drilled with slurry	$0.3c$	0.8
Belled, drilled dry	$0.3c$	0.8
Belled, drilled with slurry	$0.15c$	0.5

[1] c is soil cohesion determined from triaxial testing, not *in situ* vane shear tests.

friction (f) can be obtained from Table 8-2. The skin friction that develops along the shaft is related to the cohesion of the clay and the manner in which the caisson is drilled.

A factor of safety of 3 is recommended for Q_{bottom} [i.e., the term cN_cA_{bottom} in Eq. (8-1)]. [1] Thus,

$$Q_{allowable} = \tfrac{1}{3}cN_cA_{bottom} + fA_{shaft} \qquad (8-2)$$

EXAMPLE 8-1

Given

1. A 3-ft-diameter plain concrete drilled caisson is placed in clay soil.
2. Soil conditions and a sketch of the caisson are shown in Fig. 8-2.

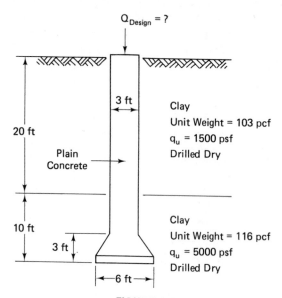

FIGURE 8-2

3. The excavation is drilled dry.

4. A local building code states: "The shafts of caissons shall be designed as concrete columns with continuous lateral support. The unit compressive stress shall not exceed 600 psi for plain nor 900 psi for reinforced concrete."

Required

Design capacity of the drilled caisson.

Solution

1. *Supporting strength of soil:*
   From Eq. (8-2),

   $$Q_{allowable} = \tfrac{1}{3}cN_cA_{bottom} + fA_{shaft} \qquad (8\text{-}2)$$

   $$c = \frac{q_u}{2} = \frac{5000}{2} = 2500 \text{ psf} = 2.5 \text{ ksf}$$

   $$\frac{\text{Depth of caisson}}{\text{Diameter of caisson bottom}} = \frac{30}{6} = 5$$

   From Table 8-1, $N_c = 9$,

   $$A_{bottom} = \frac{\pi(6)^2}{4} = 28.27 \text{ ft}^2$$

   $$\tfrac{1}{3}cN_cA_{bottom} = \frac{(2.5)(9)(28.27)}{3} = 212 \text{ kips}$$

   From Table 8-2, with belled caisson and drilled dry, adhesion or skin friction $(f) = 0.3c$ but not more than 0.8 ksf.

   $$f_1 = (0.3)\left(\frac{1500}{2}\right) = 225 \text{ psf} = 0.225 \text{ ksf} < 0.8 \text{ ksf}$$

   $$f_2 = (0.3)\left(\frac{5000}{2}\right) = 750 \text{ psf} = 0.750 \text{ ksf} < 0.8 \text{ ksf} \qquad \textbf{O.K.}$$

   A_{shaft} = circumference of caisson shaft multiplied by the effective length of shaft in developing skin friction

   $A_{shaft_1} = (\pi)(3)(20) = 188.5 \text{ ft}^2$

   $A_{shaft_2} = (\pi)(3)(10 - 3) = 66.0 \text{ ft}^2$

   $fA_{shaft} = (0.225)(188.5) + (0.750)(66.0) = 92 \text{ kips}$

   $Q_{allowable} = 212 + 92 = 304 \text{ kips}$

2. *Supporting strength of concrete shaft:*
   According to the local building code given,

   $$Q_a = 600 \times \frac{\pi(3 \times 12)^2}{4} = 610,700 \text{ lb} = 611 \text{ kips}$$

Design capacity of the caisson is the smaller of (1) and (2), or 304 kips.

Drilled Caissons in Sand

The analysis of drilled caissons in sand is also similar to that of piles, in accordance with Eq. (7-1). The term Q_{tip} of Eq. (7-1) can be evaluated by multiplying the effective vertical pressure (p_v) considering the limits imposed by the concept of critical depth by the bearing capacity factor (N_q) and this by the area of the bottom of the caisson (A_{bottom}). The term $Q_{friction}$ of Eq. (7-1) can be evaluated by multiplying the coefficient of lateral earth pressure of the soil at rest (K_0) by the effective vertical pressure (p_v) by the coefficient of friction between sand and concrete $(\tan \delta)$ by the skin area of the caisson shaft (A_{shaft}). Making these substitutions in Eq. (7-1) gives [1]

$$Q_{ultimate} = p_v N_q A_{bottom} + (K_0 p_v \tan \delta) A_{shaft} \qquad (8\text{-}3)$$

The value of the coefficient of lateral earth pressure at rest (K_0) ranges from about 0.4 for dense sand to 0.5 for loose sand [3]. The value of the coefficient of friction between sand and concrete $(\tan \delta)$ can be taken to be the value of the coefficient of friction among sand particles $(\tan \phi)$ if the excavation has been drilled dry. If the excavation is drilled using a slurry, some reduction should be applied to the value of $\tan \phi$ used [1].

When a drilled caisson in sand is designed by the procedure given above, a factor of safety of from 2 to 3 is recommended.

Example 8-2 illustrates the computation of allowable bearing capacity for a drilled caisson in sand.

EXAMPLE 8-2

Given

1. A 3-ft-diameter straight-shaft caisson is constructed in sand.
2. The soil conditions and a sketch of the caisson are shown in Fig. 8-3.
3. The excavation is drilled dry.

Required

The allowable bearing capacity of the caisson as determined by the supporting strength of the soil.

Solution

D_c = critical depth = 10 times the diameter of caisson (for loose sand)
= $10 \times 3 = 30$ ft   (see Fig. 8-4)

From Eq. (8-3),

$$Q_{ultimate} = p_v N_q A_{bottom} + (K_0 p_v \tan \delta) A_{shaft} \qquad (8\text{-}3)$$

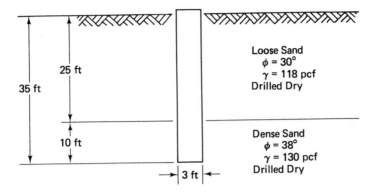

FIGURE 8-3

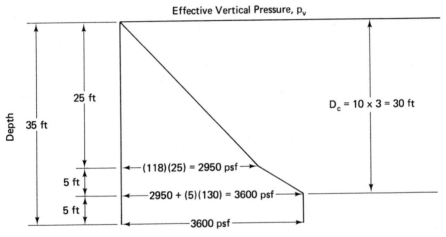

FIGURE 8-4

$p_v = 3600$ psf   (see Fig. 8-4)

$N_q = 50$ (from Fig. 6-6 for $\phi = 38°$)

$A_{bottom} = \dfrac{\pi}{4}(3)^2 = 7.07$ ft^2

$Q_{bottom} = (3600)(50)(7.07) = 1{,}273{,}000$ lb $= 1273$ kips

$Q_{friction} = (K_0)(\text{area of } p_v \text{ diagram})(\text{circumference of caisson shaft})(\tan \delta)$

$K_0 = 0.5$ (for loose sand)

$K_0 = 0.4$ (for dense sand)

$\tan \delta = \tan 30°$ (for upper layer of loose sand)

$\tan \delta = \tan 38°$ (for lower layer of dense sand)

$$Q_{friction} = (0.5)(\tfrac{1}{2} \times 2950 \times 25)(\pi \times 3)(\tan 30°)$$
$$+ (0.4)[2950 \times 5 + (\tfrac{1}{2})(5)(3600 - 2950)](\pi \times 3)(\tan 38°)$$
$$+ (0.4)(5 \times 3600)(\pi \times 3)(\tan 38°)$$
$$= 202,000 \text{ lb} = 202 \text{ kips}$$
$$Q_{ultimate} = 1273 + 202 = 1475 \text{ kips}$$
$$Q_{allowable} = \frac{1475}{3} = 492 \text{ kips}$$

Drilled Caissons on Bedrock

Clear-cut procedures do not exist for determining design capacity for drilled caissons on bedrock. The designer usually goes by the applicable local building code, which is often based on past experience in the area. Such codes may give criteria with regard to the structural strength of the concrete and/or with regard to the supporting strength of the rock. Unconfined compression tests may be performed on rock samples. Using a factor of safety of 5 to 8, the allowable bearing pressure can then be evaluated [2]. However, if allowable bearing value of rock specified by a local building code is less than the allowable bearing pressure evaluated by the results of unconfined compression tests, the allowable bearing value specified by the building code should be used.

Example 8-3 illustrates the computation of the design capacity for a drilled caisson on rock. This example problem gives a sample of a possible local building code's specification regarding drilled caissons on rock.

EXAMPLE 8-3

Given

1. A 3-ft-diameter plain concrete drilled caisson is constructed on massive limestone, the unconfined compressive strength of which is found to be 2500 psi.

2. Soil conditions and a sketch of the caisson are shown in Fig. 8-5.

3. The excavation is drilled dry.

4. A local building code states:
    a. "The shafts of caissons shall be designed as concrete columns with continuous lateral support. The unit compressive stress shall not exceed 600 psi for plain concrete nor 900 psi for reinforced concrete."
    b. "Allowable bearing value of rock shall not exceed the following:

Massive igneous or metamorphic rock	100 tons/ft²
Massive sedimentary rock	20 tons/ft²"

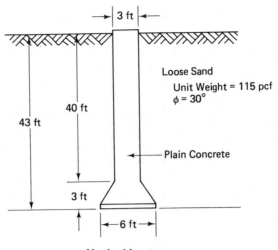

FIGURE 8-5

Required

Design capacity of the caisson (neglect skin friction of the caisson shaft).

Solution

1. The allowable bearing capacity of the caisson as determined by the structural strength of the concrete:

    $$Q_{\text{allowable}} = \frac{[(\pi)(3)^2 \text{ ft}^2](600 \text{ lb/in.}^2)(144 \text{ in.}^2/\text{ft}^2)}{4} = 611{,}000 \text{ lb}$$

    $$= 305 \text{ tons}$$

2. The allowable bearing capacity of the caisson as determined by the supporting strength of the rock:

    $$q_a = \frac{2500}{8} = 312.5 \text{ psi} = \frac{(312.5)(144)}{2000} = 22.5 \text{ tons/ft}^2$$

    (using F.S. of 8 for unconfined compressive strength of rock). However, the local building code specifies q_a shall not exceed 20 tons/ft². Therefore, q_a of 20 tons/ft² should be used.

    $$Q_{\text{allowable}} = \frac{\pi}{4}(6)^2(20) = 565 \text{ tons}$$

3. The design capacity of the caisson is the smaller of (1) and (2), or 305 tons.

8-3 SETTLEMENTS OF DRILLED CAISSONS

The settlement of drilled caissons in clay depends largely on the load history of the clay. This is similar to settlement of footings. Since drilled caissons are uneconomical in normally consolidated clay and settlement thereon is excessive, drilled caissons should be used only in overconsolidated clay. Long-term settlement analysis in clay soils can be performed using consolidation theory and assuming the bottom of the drilled caisson to be a footing [3].

The settlement of a drilled caisson in sand "at any depth is likely to be about one-half the settlement of an equally loaded footing covering the same area on sand of the same characteristics" [3]. Generally, such settlement will not be detrimental, since the caisson will normally be found on dense sand and settlement will be small. Settlement in sand can be computed using procedures given in Chap. 4 for footings on sand. It should be kept in mind, however, that the settlement of the caisson should be about one-half the settlement computed for the equivalent footing.

The settlement of drilled caissons on bedrock should be very small if the rock is dry. However, water may be found at the bottom of some caissons, and this can cause some settlement—sometimes large if soft rocks disintegrate upon soaking. It is, therefore, desirable that the water be pumped out and the caisson thoroughly cleaned during the last stage of drilling [2].

8-4 CONSTRUCTION AND INSPECTION OF DRILLED CAISSONS

The actual construction of drilled caissons consists for the most part of excavation and placement of the concrete (perhaps with reinforcing steel). As related in Sec. 8-1, drilled caissons generally are excavated using an auger drill or other type of drilling equipment. An auger is a screwlike device that is attached to a shaft and rotated under power. The rotating action digs into the soil and raises it to the surface. If a caisson is to have a bell at the bottom, the bell is made using a reamer.

While excavation is being done, the soil is exposed in the walls. Both the soil at the bottom and the soil exposed in the walls should be examined (and records kept) whenever possible to check the adequacy of the supporting soil at the bottom of the caisson and to determine the depth to, and thickness of, different soil strata. Sometimes a person may be able to descend in the shaft for inspection.

After excavation, the concrete must be of acceptable quality and properly placed. It is preferable that the concrete not strike the sides of the hole as it is being poured. A casing, if used, is generally removed as the concrete is poured.

Generally, only the concrete in the upper part of the shaft is vibrated. It is always best to pour the concrete in the dry; but if water is present, it can be be placed under water. The installation of the reinforcing steel (if specified) should be carefully checked prior to placing the concrete.

One final aspect of the overall construction process is inspection. A drilled caisson should be inspected for accuracy of the caisson (alignment and dimensions), for the bearing capacity of the soil at the bottom of the caisson, for the proper placement of the reinforcing steel and of the concrete, etc. Normally, the owner's representative should be present during construction to ensure that the construction of the caisson is done properly and according to specifications.

8-5 PROBLEMS

8-1 A plain concrete drilled caisson is to be constructed in clay soils. The diameter of the caisson shaft is 4 ft, and the belled bottom is 8 ft in diameter. The drilled caisson is extended to a total depth of 36 ft. The soil conditions are illustrated in Fig. 8-6. Compute the allowable bearing capacity of the caisson if (a) the excavation is drilled dry, and (b) the foundation is to be drilled with bentonite slurry, and a factor of safety of 3 is to be used. (*Note:* The maximum allowable compressive stress of plain concrete is assumed to be 600 psi.)

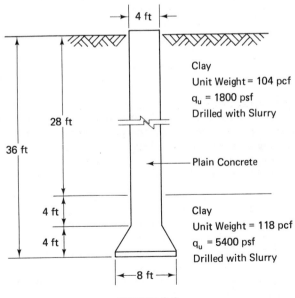

FIGURE 8-6

8-2   A drilled caisson 4 ft in diameter and supported by a bell end is to be constructed of reinforced concrete in sand. Soil conditions and a sketch of the caisson are shown in Fig. 8-7. Compute the design capacity of the caisson if the excavation is slurry drilled and the factor of safety is 3. *Note:* (1) Assume that the coefficient of friction between the sand and concrete is tan $\frac{2}{3}\phi$ for this bentonite-slurry-drilled caisson. (2) The maximum allowable compressive stress of reinforced concrete is assumed to be 900 psi.

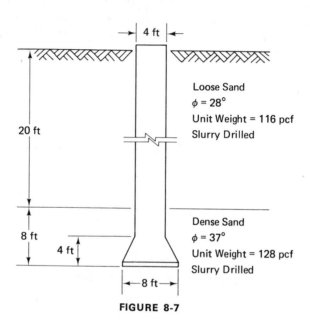

4 ft

Loose Sand
$\phi = 28°$
Unit Weight = 116 pcf
Slurry Drilled

20 ft

8 ft

4 ft

Dense Sand
$\phi = 37°$
Unit Weight = 128 pcf
Slurry Drilled

8 ft

FIGURE 8-7

8-3   A straight drilled caisson, 4 ft in diameter and made of reinforced concrete, rests on a horizontal bedded granite (massive igneous rock). The unconfined compressive strength of the intact granite sample is 20,000 psi. Determine the safe design load on the caisson if skin friction of the caisson shaft is neglected (see Fig. 8-8). *Note:* A local building code states:

1. "The unit allowable compressive stress shall not exceed 600 psi for plain concrete or 900 psi for reinforced concrete caissons."

2. "The allowable bearing value of the rock shall not exceed the following:

Massive igneous or metamorphic rock   100 tons/ft²

Massive sedimentary rock   20 tons/ft²"

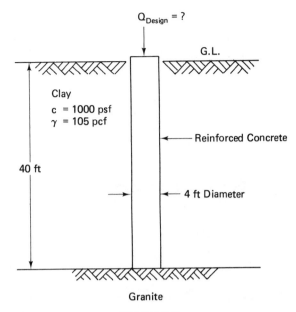

Q_{Design} = ?

G.L.

Clay
c = 1000 psf
γ = 105 pcf

Reinforced Concrete

40 ft

4 ft Diameter

Granite

FIGURE 8-8

References

[1] DAVID F. MCCARTHY, *Essentials of Soil Mechanics and Foundations*, Reston Publishing Company, Inc., Reston, Va., 1977.

[2] WAYNE C. TENG, *Foundation Design*, Prentice-Hall, Inc., Englewood Cliffs, N.J., 1962.

[3] KARL TERZAGHI AND RALPH B. PECK, *Soil Mechanics in Engineering Practice*, John Wiley & Sons, Inc., New York, 1967. Copyright © 1967, by John Wiley & Sons, Inc. Reprinted by permission of John Wiley & Sons, Inc.

9

Earth Pressure

9-1 INTRODUCTION

The word "lateral" means "to the side" or "sideways." Thus, lateral earth pressure means pressure to the side, or sideways pressure. The analysis and determination of lateral earth pressure are necessary to design retaining walls and other earth retaining structures, such as bulkheads, abutments, and the like. Obviously, the magnitude and the location of lateral earth pressure must be known in order to design a retaining wall or other retaining structure that can withstand the applied pressure with an adequate safety margin. Almost always the engineer or engineering technologist calculates earth pressures and forces on a unit (1-ft) section of the retaining wall.

There are three categories of earth pressure. These are *earth pressure at rest*, *active earth pressure*, and *passive earth pressure*. Earth pressure at rest refers to lateral pressure caused by earth that is prevented from lateral movement by an unyielding wall. In actuality, however, some retaining wall movement is almost inevitable, resulting in either active earth pressure or passive earth pressure as explained below.

If the wall moves away from the soil, as sketched in Fig. 9-1, the earth surface will tend to be lowered, and the lateral pressure on the wall will be decreased. If the wall moves far enough away, shear failure of the soil will occur, and a sliding soil wedge will tend to move forward and downward. The earth pressure exerted on the wall at this state of failure is known as the active earth pressure (P_a), and it is at the minimum value.

If, on the other hand, the wall moves toward the soil, as sketched in Fig. 9-2, the earth surface will tend to be raised, and the lateral pressure on the wall will be increased. If the wall moves far enough toward the soil, failure

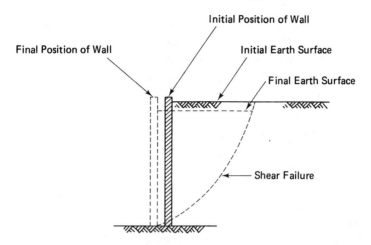

FIGURE 9-1 Active earth pressure. (For illustrative purposes, assume that the wall yields by moving outward away from the soil with its surface remaining vertical.)

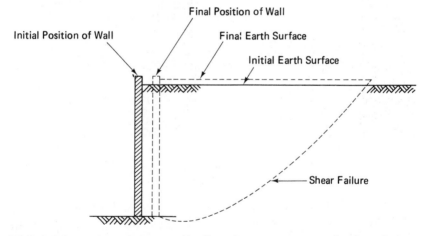

FIGURE 9-2 Passive earth pressure. (For illustrative purposes, assume that the wall moves backward toward the soil with its surface remaining vertical.)

of the soil will occur, and a sliding soil wedge will tend to move backward and upward. The earth pressure exerted on the wall at this state of failure is known as the passive earth pressure (P_p), and it is at the maximum value.

9-2 RANKINE EARTH PRESSURES

The Rankine theory for determining lateral earth pressures is based on several assumptions. The primary one is that there is no adhesion or friction between the wall and the soil (i.e., the wall is smooth). In addition, lateral pressures computed from Rankine theory are limited to vertical walls. The resultant pressure is assumed to act at a distance up from the base of the wall equal to one-third of the vertical distance from the heel at the base of the wall to the surface of the backfill (see Fig. 9-3). The direction of the resultant is parallel to the backfill surface.

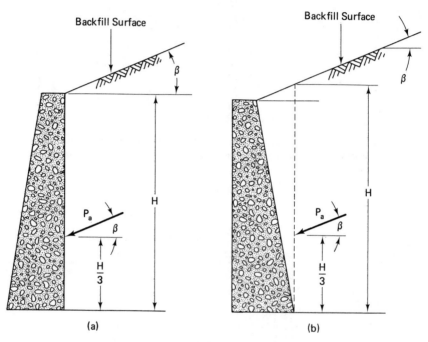

FIGURE 9-3

The primary assumption stated above (i.e., the wall is smooth) is not valid. Nevertheless, the equations derived based on this assumption are widely used for computing lateral earth pressures; and, propitiously, results obtained using these equations may not differ appreciably from results based on more accurate and sophisticated analyses. In fact, results based on Rankine theory generally give slightly larger values, causing a slightly larger wall to be designed, thus giving a slight additional safety factor.

The equations for computing lateral earth pressure[1] based on Rankine theory are as follows [1]:

$$P_a = \tfrac{1}{2}\gamma H^2 K_a \tag{9-1}$$

---

[1] P_a and P_p are actually forces per linear foot of wall; however, these are commonly referred to as the lateral earth pressure.

where

$$K_a = \cos \beta \frac{\cos \beta - \sqrt{\cos^2 \beta - \cos^2 \phi}}{\cos \beta + \sqrt{\cos^2 \beta - \cos^2 \phi}} \qquad (9\text{-}2)$$

$$P_p = \tfrac{1}{2}\gamma H^2 K_p \qquad (9\text{-}3)$$

where

$$K_p = \cos \beta \frac{\cos \beta + \sqrt{\cos^2 \beta - \cos^2 \phi}}{\cos \beta - \sqrt{\cos^2 \beta - \cos^2 \phi}} \qquad (9\text{-}4)$$

where P_a = active earth pressure

γ = unit weight of the backfill soil

H = height of the wall (see Fig. 9-3)

K_a = coefficient of active earth pressure

β = angle between backfill surface line and a horizontal line (see Fig. 9-3)

ϕ = angle of internal friction of the backfill soil

P_p = passive earth pressure

K_p = coefficient of passive earth pressure

If the backfill surface is level, the angle β is zero, and Eqs. (9-2) and (9-4) revert to

$$K_a = \frac{1 - \sin \phi}{1 + \sin \phi} \qquad (9\text{-}5)$$

$$K_p = \frac{1 + \sin \phi}{1 - \sin \phi} \qquad (9\text{-}6)$$

Example 9-1 illustrates the computation of lateral earth pressure for a level backfill surface, and Example 9-2 illustrates the computation for a sloping backfill surface. Example 9-3 illustrates a technique for computing lateral earth pressure (based on Rankine theory) for a retaining wall with a back side that is not vertical.

EXAMPLE 9-1

Given

A retaining wall as shown in Fig. 9-4.

Required

The total active earth pressure per foot of wall and the point of application of the total earth pressure by Rankine theory.

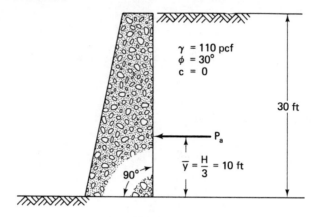

FIGURE 9-4

Solution

From Eqs. (9-1) and (9-5) (for level backfill),

$$P_a = \tfrac{1}{2}\gamma H^2 K_a \tag{9-1}$$

$$K_a = \frac{1 - \sin \phi}{1 + \sin \phi} \tag{9-5}$$

$$K_a = \frac{1 - \sin 30°}{1 + \sin 30°} = 0.333$$

$$P_a = (\tfrac{1}{2})(110)(30)^2(0.333) = 16{,}500 \text{ lb/ft}$$

Point of application of the total earth pressure $(\bar{y}) = H/3 = 30/3 = 10$ ft from the base of the wall.

EXAMPLE 9-2

Given

A retaining wall as shown in Fig. 9-5.

Required

The total active earth pressure per foot of wall and the location of the pressure by Rankine theory.

Solution

From Eqs. (9-1) and (9-2),

$$P_a = \tfrac{1}{2}\gamma H^2 K_a \tag{9-1}$$

$$K_a = \cos \beta \frac{\cos \beta - \sqrt{\cos^2 \beta - \cos^2 \phi}}{\cos \beta + \sqrt{\cos^2 \beta - \cos^2 \phi}} \tag{9-2}$$

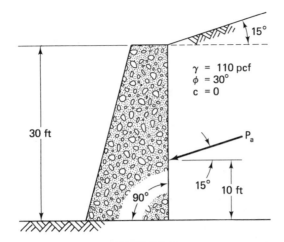

FIGURE 9-5

$$K_a = (\cos 15°) \frac{\cos 15° - \sqrt{\cos^2 15° - \cos^2 30°}}{\cos 15° + \sqrt{\cos^2 15° - \cos^2 30°}} = 0.373$$

$$P_a = (\tfrac{1}{2})(110)(30)^2(0.373) = 18,500 \text{ lb/ft}$$

$\bar{y} = H/3 = 30/3 = 10$ ft from the base of the wall (see Fig. 9-5).

EXAMPLE 9-3

Given

A retaining wall as shown in Fig. 9-6.

Required

The total active earth pressure per foot of wall by Rankine theory.

Solution

As shown on Fig. 9-6,

$$\tan 5° = \frac{AB}{20}$$

$$AB = (20)(\tan 5°) = 1.75 \text{ ft}$$

Also,

$$\tan 10° = \frac{Og}{AB} = \frac{h}{1.75}$$

$$h = (1.75)(\tan 10°) = 0.31 \text{ ft}$$

From Eqs. (9-1) and (9-2),

$$P_a' = \tfrac{1}{2}\gamma H^2 K_a \qquad\qquad (9\text{-}1)$$

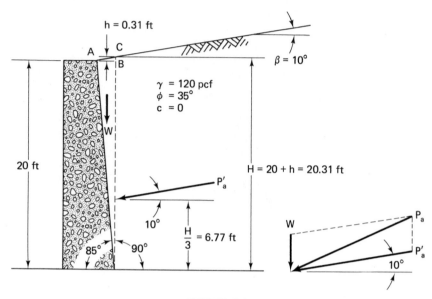

FIGURE 9-6

$$K_a = \cos \beta \frac{\cos \beta - \sqrt{\cos^2 \beta - \cos^2 \phi}}{\cos \beta + \sqrt{\cos^2 \beta - \cos^2 \phi}} \qquad (9\text{-}2)$$

$\gamma = 120$ pcf

$H = 20.31$ ft

$\beta = 10°$

$\phi = 35°$

$$K_a = (\cos 10°) \frac{\cos 10° - \sqrt{\cos^2 10° - \cos^2 35°}}{\cos 10° + \sqrt{\cos^2 10° - \cos^2 35°}} = 0.282$$

$P'_a = (\frac{1}{2})(120)(20.31)^2(0.282) = 6979$ lb/ft

$W = (\frac{1}{2})(\gamma)(AB)(H)$

$W = \frac{1}{2}(120)(1.75)(20.31) = 2133$ lb/ft

$P_h = P'_a \cos \beta = (6979) \cos 10° = 6873$ lb/ft

$P_v = P'_a \sin \beta = (6979) \sin 10° = 1212$ lb/ft

$\sum V = W + P_v = 2133 + 1212 = 3345$ lb/ft

$\sum H = P_h = 6873$ lb/ft

Total active earth pressure $(P_a) = \sqrt{(\sum V)^2 + (\sum H)^2} = \sqrt{(3345)^2 + (6873)^2}$
$= 7640$ lb/ft.

Equations (9-1) through (9-6) are applicable in the case of cohesionless
soils. The generalized lateral earth pressure distribution for soils that have

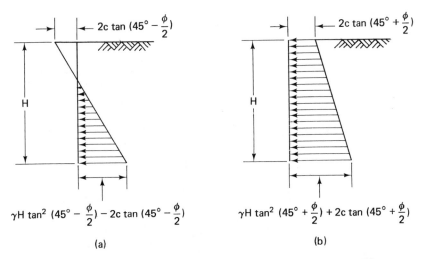

FIGURE 9-7 (a) Active earth pressure; (b) passive earth pressure. [2]

both cohesion and friction is, based on Rankine theory, as shown in Fig. 9-7. Figure 9-7a gives the pressure distribution for active pressure, and Fig. 9-7b gives that for passive pressure. It will be noted that the active pressure acts only over the lower part of the wall (see Fig. 9-7a). The pressure distribution for a particular case can be ascertained by substituting the appropriate parameters into the equations indicated on Fig. 9-7. Example 9-4 illustrates this method.

EXAMPLE 9-4

Given

The retaining wall shown in Fig. 9-8.

Required

Draw the active earth pressure diagram by Rankine theory.

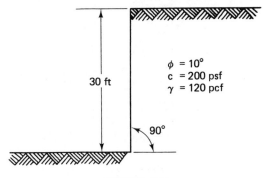

FIGURE 9-8

Solution

From Fig. 9-7a,

$$2c \tan\left(45° - \frac{\phi}{2}\right) = (2)(200) \tan\left(45° - \frac{10°}{2}\right)$$

$$= 336 \text{ psf}$$

$$\gamma H \tan^2\left(45° - \frac{\phi}{2}\right) - 2c \tan\left(45° - \frac{\phi}{2}\right) = (120)(30) \tan^2\left(45° - \frac{10°}{2}\right)$$

$$- (2)(200) \tan\left(45° - \frac{10°}{2}\right) = 2200 \text{ psf}$$

Thus, the pressure distribution is as shown in Fig. 9-9. By direct proportion of two triangles,

$$\frac{366}{2200} = \frac{30 - x}{x}$$

$$366x = (2200)(30 - x)$$

$$366x = 66,000 - 2200x$$

$$2566x = 66,000$$

$$x = 25.72 \text{ ft}$$

Resultant $= \frac{1}{2}(2200)(25.72) = 28,300$ lb/ft
$\bar{y} = x/3 = 25.72/3 = 8.57$ ft above the base of the wall.

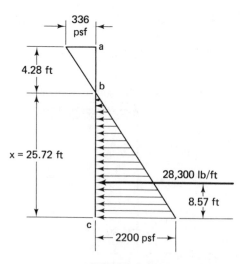

FIGURE 9-9

9-3 COULOMB EARTH PRESSURES

The Coulomb theory for determining lateral earth pressure, developed nearly a century before the Rankine theory, assumes that failure occurs in the form of a wedge and that friction occurs between the wall and the soil. The sides of the wedge are the earth side of the retaining wall and a failure plane that passes through the toe of the wall (see Fig. 9-10).

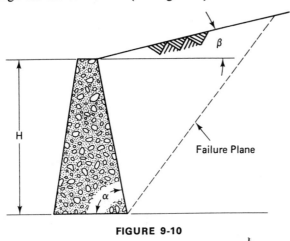

FIGURE 9-10

The resultant active earth pressure acts on the wall at a point where a line through the center of gravity of the wedge and parallel to the failure plane intersects the wall (see Fig. 9-11). The direction of the resultant at the wall is along a line that makes an angle δ with a line normal to the back side of the wall, where δ is the angle of wall friction (see Fig. 9-12).

The equations for computing lateral earth pressure based on Coulomb theory are as follows [1]:

$$P_a = \tfrac{1}{2}\gamma H^2 K_a \tag{9-1}$$

where

$$K_a = \frac{\sin^2(\alpha + \phi)}{\sin^2\alpha \sin(\alpha - \delta)\left[1 + \sqrt{\dfrac{\sin(\phi + \delta)\sin(\phi - \beta)}{\sin(\alpha - \delta)\sin(\alpha + \beta)}}\right]^2} \tag{9-7}$$

$$P_p = \tfrac{1}{2}\gamma H^2 K_p \tag{9-3}$$

where

$$K_p = \frac{\sin^2(\alpha - \phi)}{\sin^2\alpha \sin(\alpha + \delta)\left[1 - \sqrt{\dfrac{\sin(\phi + \delta)\sin(\phi + \beta)}{\sin(\alpha + \delta)\sin(\alpha + \beta)}}\right]^2} \tag{9-8}$$

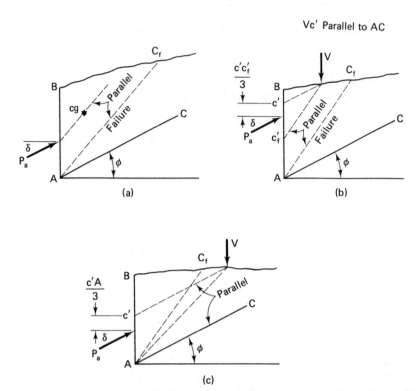

FIGURE 9-11 Procedures for location of point of application of P_a: (a) irregular backfill; (b) concentrated or line load inside failure zone; (c) concentrated or line load outside failure zone (but inside zone ABC). [1]

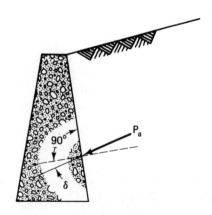

FIGURE 9-12

where P_a = active earth pressure

   γ = unit weight of the backfill soil

   H = height of the retaining wall (see Fig. 9-10)

   K_a = coefficient of active earth pressure

   α = angle between back side of wall and a horizontal line (see Fig. 9-10)

   ϕ = angle of internal friction of the backfill soil

   δ = angle of wall friction

   β = angle between backfill surface line and a horizontal line (see Fig. 9-10)

   P_p = passive earth pressure

   K_p = coefficient of passive earth pressure

In the case of a smooth, vertical wall with level backfill, δ and β are each zero and α is 90°; and if these values are substituted into Eqs. (9-7) and (9-8), the equations revert to Eqs. (9-5) and (9-6), respectively. The latter two equations are the Rankine equations for the conditions stated (i.e., smooth, vertical wall with level backfill).

Table 9-1 gives some typical values of angles of internal friction, angles of wall friction, and unit weights of common types of backfill soil.

Examples 9-5 and 9-6 illustrate the use of the equations based on Coulomb theory.

EXAMPLE 9-5

Given

Same conditions as in Example 9-1, except that the angle of wall friction between backfill and wall (δ) is 25° (see Fig. 9-13).

Required

The total active earth pressure per foot of wall by Coulomb theory.

Solution

From Eqs. (9-1) and (9-7),

$$P_a = \tfrac{1}{2}\gamma H^2 K_a \tag{9-1}$$

$$K_a = \frac{\sin^2 (\alpha + \phi)}{\sin^2 \alpha \sin (\alpha - \delta) \left[1 + \sqrt{\dfrac{\sin (\phi + \delta) \sin (\phi - \beta)}{\sin (\alpha - \delta) \sin (\alpha + \beta)}}\right]^2} \tag{9-7}$$

$$\gamma = 110 \text{ pcf}$$

$$H = 30 \text{ ft}$$

TABLE 9-1   Friction angles and unit weights for backfill soil [3].

Number	Description of Soil	Angle of Internal Friction, ϕ		Angle of Wall Friction, δ		Unit Weight, γ (pcf)	
		Dry	Moist	Dry	Moist	Dry	Moist
1	Coarse to medium sand, trace fine gravel	36°00′	27°30′	27°30′	26°10′	—	91
2	Coarse to fine sand, trace + silt (7.5%)	37°40′	27°50′	32°10′	26°20′	101	95
3	Coarse to fine sand, trace + (7.5%) fine gravel	38°40′	30°00′	27°10′	26°20′	106	94
4	Coarse to fine sand	36°30′	30°00′	28°50′	27°10′	95	80
5	Medium to fine sand, some silt (29%), trace fine gravel	35°10′	29°10′	25°10′	21°30′	99	82
6	Fine sand, trace silt	37°50′	29°20′	29°40′	26°20′	94	82
7	Fine sand, some silt	35°00′	30°20′	28°00′	28°00′	103	96
8	Coarse silt, fine sand (45%)	34°50′	26°10′	27°50′	25°40′	94	80
9	Silt, some coarse to fine sand, trace + clay (7%)	—	31°20′	—	28°50′	—	75

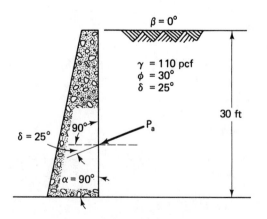

FIGURE 9-13

$$\alpha = 90°$$
$$\phi = 30°$$
$$\delta = 25°$$
$$\beta = 0° \quad \text{(level backfill)}$$

$$K_a = \frac{\sin^2 (90° + 30°)}{\sin^2 (90°) \sin (90° - 25°) \left[1 + \sqrt{\dfrac{\sin (30° + 25°) \sin (30° - 0°)}{\sin (90° - 25°) \sin (90° + 0°)}} \right]^2}$$

$$= 0.296$$
$$P_a = (\tfrac{1}{2})(110)(30)^2(0.296) = 14,700 \text{ lb/ft}$$

EXAMPLE 9-6

Given

Same conditions as Example 9-3 except that the angle of wall friction between backfill and wall (δ) is 20° (see Fig. 9-14).

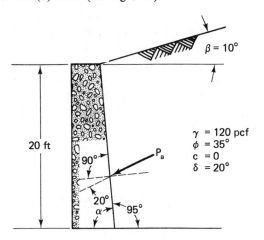

$\beta = 10°$

20 ft

90°

P_a

$\gamma = 120$ pcf
$\phi = 35°$
$c = 0$
$\delta = 20°$

20°
α 95°

FIGURE 9-14

Required

The total active earth pressure per foot of wall by Coulomb theory.

Solution

From Eqs. (9-1) and (9-7),

$$P_a = \tfrac{1}{2}\gamma H^2 K_a \tag{9-1}$$

$$K_a = \frac{\sin^2 (\alpha + \phi)}{\sin^2 \alpha \sin (\alpha - \delta) \left[1 + \sqrt{\dfrac{\sin (\phi + \delta) \sin (\phi - \beta)}{\sin (\alpha - \delta) \sin (\alpha + \beta)}} \right]^2} \tag{9-7}$$

$$\gamma = 120 \text{ pcf}$$

$$H = 20 \text{ ft}$$

$$\alpha = 180° - 95° = 85°$$

$$\phi = 35°$$

$$\delta = 20°$$

$$\beta = 10°$$

$$K_a = \frac{\sin^2(85° + 35°)}{\sin^2(85°)\sin(85° - 20°)\left[1 + \sqrt{\dfrac{\sin(35° + 20°)\sin(35° - 10°)}{\sin(85° - 20°)\sin(85° + 10°)}}\right]^2}$$

$$= 0.318$$

$$P_a = (\tfrac{1}{2})(120)(20)^2(0.318) = 7630 \text{ lb/ft}$$

9-4 EFFECTS OF SURCHARGE LOAD UPON ACTIVE THRUST

Sometimes the backfill resting against a retaining wall is subjected to a surcharge. A surcharge, which is simply a uniform load and/or concentrated load imposed on the soil, adds to the lateral earth pressure exerted against the retaining wall by the backfill. This added pressure must, of course, be considered when designing the retaining wall.

The additional pressure exerted against a retaining wall as a result of a surcharge in the form of a uniform load can be computed from the following equation (see Fig. 9-15) [4]:[1]

$$P' = qHK_a \tag{9-9}$$

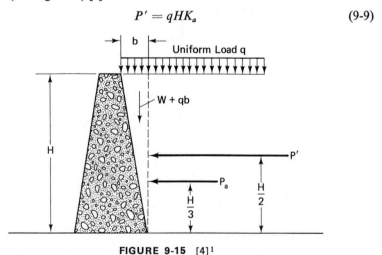

FIGURE 9-15 [4][1]

[1] Reprinted with permission of Macmillan Publishing Co., Inc., from *Theory and Practice of Foundation Engineering* by Louis J. Goodman and R. H. Karol. Copyright © 1968, Macmillan Publishing Co., Inc.

where $P' =$ additional active earth pressure as a result of uniform load surcharge

   $q =$ uniform load (surcharge) on backfill

   $H =$ height of wall

   $K_a =$ coefficient of active earth pressure [determined from Eq. (9-5)]

Example 9-7, which follows, illustrates the computation of pressure due to a surcharge in the form of a uniform load. Example 9-10 in Sec. 9-5 illustrates the treatment (graphical solution) of a surcharge in the form of a concentrated load.

EXAMPLE   9-7

Given

1. A smooth vertical wall is 20 ft high and retains a cohesionless soil with $\gamma = 120$ pcf and $\phi = 28°$.

2. The top of the soil is level with the top of the wall and is horizontal.

3. The soil surface carries a uniformly distributed load of 1000 lb/ft² (see Fig. 9-16).

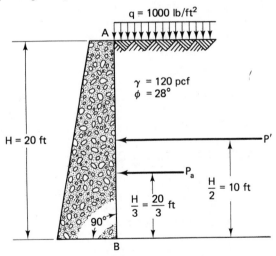

FIGURE 9-16

Required

1. The total active earth pressure on the wall per linear foot of wall.

2. The point of action of the total active earth pressure by Rankine theory.

Solution

From Eqs. (9-1) and (9-5) (for level backfill),

$$P_a = \tfrac{1}{2}\gamma H^2 K_a \tag{9-1}$$

$$K_a = \frac{1 - \sin \phi}{1 + \sin \phi} \tag{9-5}$$

$$K_a = \frac{1 - \sin 28°}{1 + \sin 28°} = 0.361$$

$$P_a = (\tfrac{1}{2})(120)(20)^2(0.361) = 8660 \text{ lb/ft}$$

Point of action for $P_a = H/3 = 20/3$ ft from the base of the wall.
From Eq. (9-9),

$$P' = qHK_a \tag{9-9}$$

$$P' = (1000)(20)(0.361) = 7220 \text{ lb/ft}$$

Point of action for $P' = H/2 = 20/2 = 10$ ft from the base of the wall.

1. Total active earth pressure $= P_a + P' = 8660 + 7220 = 15,880$ lb/ft.
2. Let the point of application of the total active earth pressure be "h" ft above the base of the wall. "h" is obtained by taking moments of forces (i.e., P_a and P') at the base of the wall.

$$(15,880)(h) = (8660)\left(\frac{20}{3}\right) + (7220)(10)$$

$$h = 8.18 \text{ ft}$$

The total active earth pressure acts at 8.18 ft above the base of the wall.

9-5 CULMANN'S GRAPHICAL SOLUTION

Several graphical methods to determine earth pressures are available, one of which is the Culmann's graphical solution. The steps in carrying out a Culmann's graphical solution for active earth pressure (P_a) may be summarized as follows [1]:

1. Draw the retaining wall, backfill, and so on, to a convenient scale (see Fig. 9-17).
2. From point A (the base of the wall), lay off a line at an angle ϕ (angle of internal friction) with a horizontal line. This is line AC in Fig. 9-17.

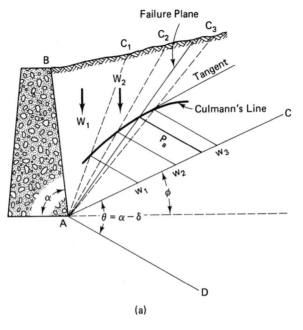

(a)

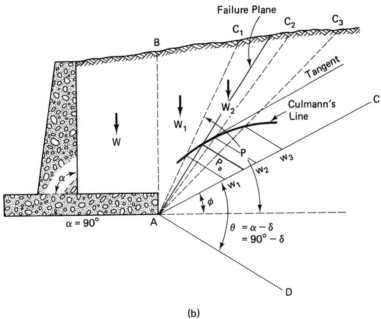

(b)

FIGURE 9-17 [1]

245

3. From point A, lay off a line at an angle θ with line AC (step 2). The angle θ is equal to α (the angle between the back side of the wall and a horizontal line, as indicated in Fig. 9-17) minus δ (angle of wall friction). This line is AD in Fig. 9-17.

4. Draw some possible failure wedges, such as ABC_1, ABC_2, ABC_3, and so on.

5. Compute the weights of the wedges (W_1, W_2, W_3, etc.).

6. Using a convenient weight scale along line AC, lay off the respective weights of the wedges, locating point w_1, w_2, w_3, and so on.

7. Through each point, w_1, w_2, w_3, and so on, draw a line parallel to line AD, intersecting the corresponding line AC_1, AC_2, AC_3, respectively.

8. Draw a smooth curve (*Culmann's line*) through the points of intersection determined in step 7 (i.e., the point of intersection of the line through point w_1 parallel to line AD and of the line AC_1, the point of intersection of the line through point w_2 parallel to line AD and of the line AC_2, etc.)

9. Draw a line that is both tangent to the Culmann line and parallel to line AC.

10. Draw a line through the tangent point (determined in step 9) that is parallel to line AD and intersects line AC. The length of this line applied to the weight scale gives the value of P_a (see Fig. 9-17). A line from point A through the tangent point defines the failure plane.

As discussed in Sec. 9-3, the point of application of P_a can be found by drawing a line through the center of gravity of the failure wedge and parallel to the failure plane until it intersects the wall (see Fig. 9-11). The direction of P_a is along a line that makes an angle δ (δ is the angle of wall friction) with a line normal to the back side of the wall (see Fig. 9-12).

Examples 9-8, 9-9, and 9-10 illustrate the application of Culmann's graphical solution.

EXAMPLE 9-8

Given

The same conditions as in Example 9-6 (see Fig. 9-18).

Required

The total active earth pressure per foot of wall by Culmann's graphical solution.

Solution

Following the steps outlined previously for Culmann's graphical solution, the sketch of Fig. 9-19 is prepared. The weights of the wedges (step 5) are

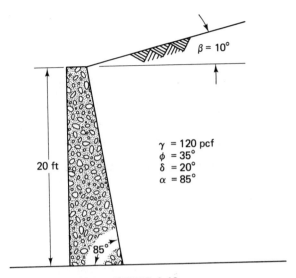

γ = 120 pcf
ϕ = 35°
δ = 20°
α = 85°

FIGURE 9-18

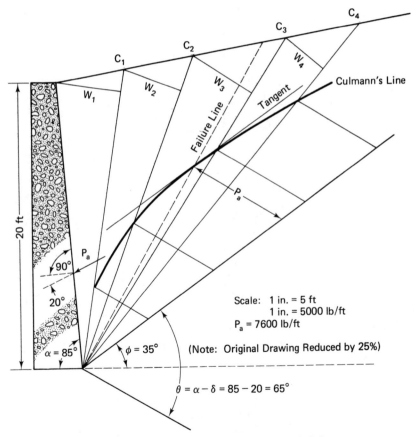

Scale: 1 in. = 5 ft
        1 in. = 5000 lb/ft
P_a = 7600 lb/ft

(Note: Original Drawing Reduced by 25%)

$\theta = \alpha - \delta = 85 - 20 = 65°$

FIGURE 9-19 Culmann's solution for Example 9-8.

computed as follows:

$$w_1 = (\tfrac{1}{2})(120)(4.7)(21) = 5920 \text{ lb/ft}$$
$$w_2 = (\tfrac{1}{2})(120)(4.4)(22.2) = 5860 \text{ lb/ft}$$
$$w_3 = (\tfrac{1}{2})(120)(5.0)(27.2) = 8160 \text{ lb/ft}$$
$$w_4 = (\tfrac{1}{2})(120)(3.5)(31.4) = 6590 \text{ lb/ft}$$

From Fig. 9-19, the value of P_a is determined to be 7600 lb/ft.

EXAMPLE 9-9

Given

Same conditions as in Example 9-7 (see Fig. 9-20).

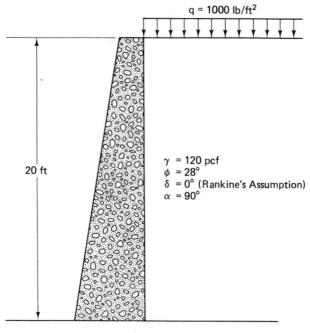

q = 1000 lb/ft²

20 ft

γ = 120 pcf
ϕ = 28°
δ = 0° (Rankine's Assumption)
α = 90°

FIGURE 9-20

Required

The total active earth pressure per foot of wall by Culmann's graphical solution.

Solution

The effect of the surcharge uniform load $q = 1000$ lb/ft², such as highway loading, is taken into account by superposing an equivalent depth of fill $h = q/\gamma = 1000/120 = 8.33$ ft on each trial wedge. Then, the Culmann's graphical solution is carried out by following the steps outlined previously and preparing the sketch of Fig. 9-21. The weights of the wedges (step 5) are computed as follows:

$$w_1 = (\tfrac{1}{2})(120)(5)(20) + (120)(8.33)(5) = 11,000 \text{ lb/ft}$$
$$w_2 = (\tfrac{1}{2})(120)(4.5)(22.4) + (120)(8.33)(5) = 11,050 \text{ lb/ft}$$
$$w_3 = (\tfrac{1}{2})(120)(4.0)(25.0) + (120)(8.33)(5) = 11,000 \text{ lb/ft}$$
$$w_4 = (\tfrac{1}{2})(120)(3.5)(28.3) + (120)(8.33)(5) = 10,940 \text{ lb/ft}$$

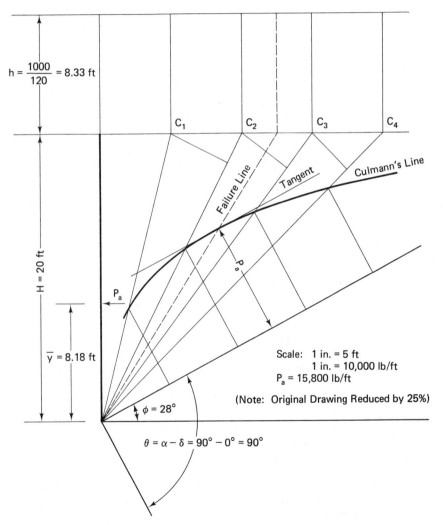

FIGURE 9-21 Culmann's solution for Example 9-9.

From Fig. 9-21 the value of P_a is determined to be 15,800 lb/ft. As computed in Example 9-7, P_a acts 8.18 ft from the base of the wall (see Fig. 9-21).

EXAMPLE 9-10

Given

1. The retaining wall shown in Fig. 9-22.
2. Unit weight of the backfill material is 120 pcf.
3. Angle of internal friction of the backfill material is 30°.

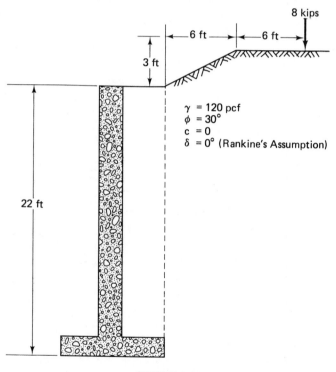

FIGURE 9-22

Required

The total active earth pressure, P_a, by Culmann's graphical solution.

Solution

Following the steps outlined previously for Culmann's graphical solution, the sketch of Fig. 9-23 is prepared. The weights of the wedges (step 5) are computed as follows:

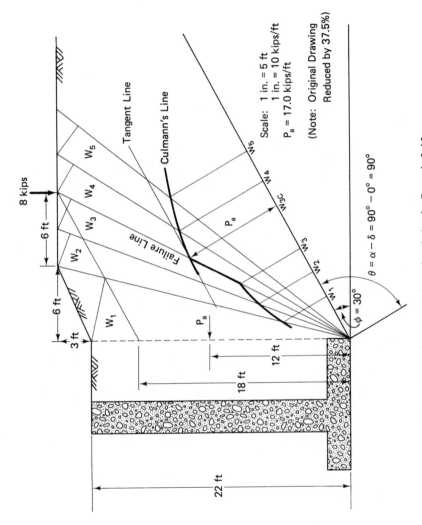

8 kips

6 ft

6 ft

3 ft

W_1

W_2

W_3

W_4

W_5

Tangent Line

Culmann's Line

Failure Line

P_a

P_a

12 ft

18 ft

22 ft

W_1
W_2
W_3
W_3C
W_4
W_5

Scale: 1 in. = 5 ft
1 in. = 10 kips/ft
P_a = 17.0 kips/ft

(Note: Original Drawing
Reduced by 37.5%)

$\theta = \alpha - \delta = 90° - 0° = 90°$

$\phi = 30°$

FIGURE 9-23 Culmann's solution for Example 9-10.

251

$$w_1 = (\tfrac{1}{2})(0.12)(5.2)(25.5) = 7.96 \text{ kips/ft}$$
$$w_2 = (\tfrac{1}{2})(0.12)(2.9)(26.4) = 4.59 \text{ kips/ft}$$
$$w_3 = (\tfrac{1}{2})(0.12)(2.7)(27.6) = 4.47 \text{ kips/ft}$$
$$w_{3c} = 8 \text{ kips} \quad (\text{concentrated load})$$
$$w_4 = (\tfrac{1}{2})(0.12)(2.6)(29.0) = 4.52 \text{ kips/ft}$$
$$w_5 = (\tfrac{1}{2})(0.12)(3.0)(31.0) = 5.58 \text{ kips/ft}$$

From Fig. 9-23, the value of P_a is determined to be 17.0 kips/ft.

9-6 DESIGN CONSIDERATIONS FOR RETAINING WALLS

In designing retaining walls, the first step is to determine the magnitude and location of the active earth pressures that will be acting on the wall. These determinations can be made by utilizing any of the methods presented previously in this chapter. Active earth pressure is used to design free-standing retaining walls.

The next step is to assume a retaining wall size. Normally, the required height of the wall will be known, and thus a wall thickness and width of the base of the wall must be estimated. The assumed wall must then be checked for three conditions. First, the wall must be safe against sliding horizontally. Second, the wall must be safe against overturning. Third, the wall must not introduce a contact pressure on the foundation soil beneath the base of the wall that exceeds the allowable bearing pressure of the foundation soil. If any of these conditions is not safe, the assumed size of the wall must be modified, and the conditions checked again. If (when) the three conditions are met, the assumed size is used for design. If, however, the three conditions are met with plenty to spare, the size might be reduced somewhat, and the conditions checked again. Obviously, this is more or less a trial-and-error procedure.

The preceding gives a brief preview of design considerations for retaining walls. This topic will be addressed in greater detail in Chap. 10.

9-7 PROBLEMS

9-1 A vertical retaining wall 25 ft high supports a deposit of sand having a level backfill. The soil properties are as follows:

$$\gamma = 120 \text{ pcf}$$
$$\phi = 35°$$
$$c = 0$$

Calculate the total active earth pressure per foot of wall and the point of application of this total earth pressure by Rankine theory.

9-2 A vertical retaining wall 25 ft high supports a deposit of sand with a sloping backfill. The angle of sloping backfill is 10°. The soil properties are as follows:

$$\gamma = 120 \text{ pcf}$$
$$\phi = 35°$$
$$c = 0$$

Calculate the total active earth pressure per foot of wall and the location of the lateral earth pressure by Rankine theory.

9-3 What is the total active earth pressure per foot of wall for the soil–wall system shown in Fig. 9-24, using Rankine theory?

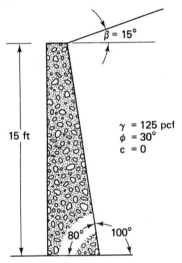

$\beta = 15°$

15 ft

$\gamma = 125$ pcf
$\phi = 30°$
$c = 0$

80°   100°

FIGURE 9-24

9-4 A vertical wall 25 ft high supports a level backfill of clayey sand. The samples of the backfill soil were tested in the laboratory and the following properties were determined: $\phi = 20°$, $c = 250$ psf, and $\gamma = 125$ pcf. Draw the active earth pressure diagram using Rankine theory.

9-5 What is the total active earth pressure per foot of wall for the retaining wall in Problem 9-1, with angle of wall friction between backfill and wall (δ) of 20°, using Coulomb theory?

9-6 What is the total earth pressure per foot of wall for the retaining wall shown in Problem 9-3? Assume an angle of wall friction between backfill and wall of 25° and use Coulomb theory.

9-7 A smooth vertical wall is 25 ft high and retains a cohesionless soil with $\gamma = 115$ pcf and $\phi = 30°$. The top of the soil is level with the top of the wall and the soil surface carries a uniformly distributed load of 500 lb/ft². Calculate the total active earth pressure on the wall per linear foot of wall, and determine the location of the total active earth pressure by Rankine theory.

9-8 Solve Problem 9-6 by Culmann's graphical solution.

9-9 Solve Problem 9-7 by Culmann's graphical solution.

References

[1] JOSEPH E. BOWLES, *Foundation Analysis and Design*, McGraw-Hill Book Company, New York, 1968.

[2] WAYNE C. TENG, *Foundation Design*, Prentice-Hall, Inc., Englewood Cliffs, N.J., 1962.

[3] WILLIAM S. LALONDE, JR., AND MILO F. JANES, eds., *Concrete Engineering Handbook*, McGraw-Hill Book Company, New York, 1961.

[4] LOUIS J. GOODMAN AND R. H. KAROL, *Theory and Practice of Foundation Engineering*, Macmillan Publishing Co., Inc., New York, 1968.

10
Retaining Walls

10-1 INTRODUCTION

A retaining wall is a wall, often made of concrete, built for the purpose of retaining, or holding back, a soil mass (or other material). A simple retaining wall is illustrated in Fig. 10-1. This type of wall depends on its weight to achieve its stability; hence, it is called a *gravity wall*. In the case of taller walls, the large lateral pressure tends to overturn the wall, and for economic reasons *cantilever walls* may be more desirable. As illustrated in Fig. 10-2, a cantilever wall has part of the base extending underneath the backfill, and (as will be shown subsequently) the weight of the soil above this part of the base helps prevent overturning.

The gravity wall is often built of plain concrete and is bulky. The concrete cantilever wall is generally more slender and must be adequately reinforced with steel. Although there are additional types of retaining walls, these two types are most common.

Although retaining walls may give the appearance of being unyielding, some forward movement of the wall is to be expected. In order that walls may undergo some forward yielding without appearing to tip over, they are often built with an inward slope on the outer face of the wall, as shown in Figs. 10-1 and 10-2. This inward slope is called *batter*.

The material placed behind a retaining wall is commonly referred to as *backfill*. It is highly desirable that backfill be a select, free-draining, granular material, such as clean sand, gravel, or broken stones. If necessary, such material should be hauled in from an area outside of the construction site. Clayey soils make extremely objectionable backfill material because of excessive

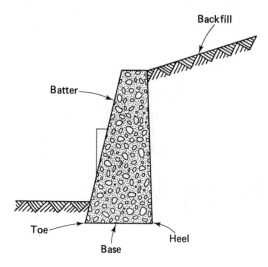

FIGURE 10-1

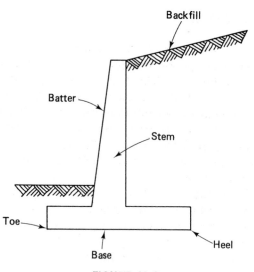

FIGURE 10-2

lateral pressure that they create. The designer of a retaining wall should either (1) write the specifications for the backfill, and base the design of the wall on the specified backfill; or (2) be given information on the material to be used as backfill and base the design of the wall on the indicated backfill. If it is possible that the water table may rise in the backfill, special designing, construction, and monitoring must go into effect.

In Chap. 9, several methods were presented for analyzing both the magnitude and the location of the lateral earth pressure acting on retaining walls. For economic reasons, retaining walls are commonly designed for active earth pressure, developed by a free-draining, granular backfill acting on the wall. As related near the end of the last chapter, a retaining wall (1) must be able to resist sliding along the base, (2) must be able to resist overturning, and (3) must not introduce a contact pressure on the foundation soil beneath the base of the wall that exceeds the allowable bearing pressure of the foundation soil. (Walls must also meet structural requirements, such as shear and bending moment; however, such considerations are not covered in this book.) Chapter 10 deals in more detail with retaining wall design.

10-2 EARTH PRESSURE COMPUTATION

In order to design a retaining wall, it is, of course, necessary to determine the earth pressure acting on the wall. Analytical determinations of such earth pressures—including Rankine earth pressure, Coulomb earth pressure, and Culmann's graphical solution—were covered in detail in Chap. 9. Retaining wall design is normally based on active earth pressure.

In practice, earth pressures for walls less than 20 ft high may be obtained from graphs or tables. Almost all such graphs and tables are developed from Rankine theory. One such graphical relationship is given in Fig. 10-3. The use of this approach to obtain earth pressure should be self-explanatory for the reader.

As will be noted by both the analytical methods of Chap. 9 and the graphical method of Fig. 10-3, the magnitude of the earth pressure on a retaining wall depends in part upon the type of soil backfill.

10-3 STABILITY ANALYSIS

The common procedure in retaining wall design is to assume a trial wall shape and size and then to check the trial wall for stability. If it does not prove to be stable by conventional standards, the shape and/or size of the wall must be revised and the revised wall must be checked for stability. This procedure is repeated until a satisfactory wall is found.

If a wall is *stable*, it means, of course, that the wall does not move. Essentially, there are three means by which a retaining wall can move—horizontally (by sliding), vertically (by excessive settlement and/or bearing capacity failure of the foundation soil), and by rotation (by overturning). The standard procedure is to check for stability with respect to each of the three means of movement to ensure that an adequate factor of safety is present in each case. The checks for sliding and for overturning hark back to the basic laws of

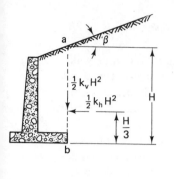

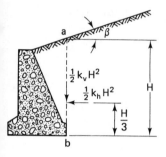

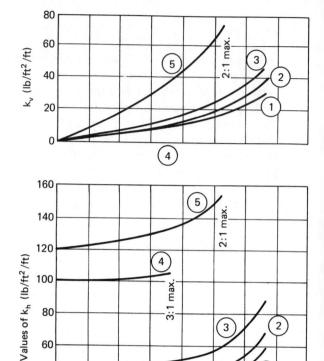

Notes:

Numerals on curves indicate soil types as described below

For material of Type 5, computations should be based on value of H 4 ft less than actual value.

Types of Backfill for Retaining Walls

(1) Coarse-grained soil without admixture of fine soil particles, very free-draining (clean sand, gravel or broken stone)

(2) Coarse-grained soil of low permeability, owing to admixture of particles of silt size

(3) Fine silty sand; granular materials with conspicuous clay content; or residual soil with stones

(4) Soft or very soft clay; organic silt; or soft silty clay

(5) Medium or stiff clay that may be placed in such a way that a negligible amount of water will enter the spaces between the chunks during floods or heavy rains

FIGURE 10-3 Earth pressure charts for retaining walls less than 20 ft high. [1, 2]

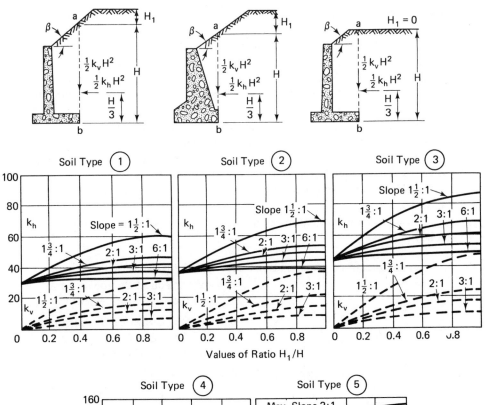

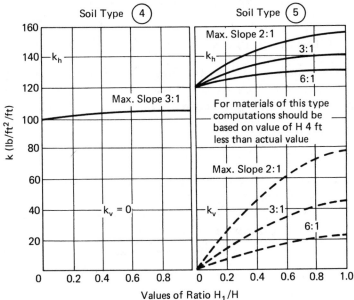

FIGURE 10-3 (Continued)

statics. The checks for settlement and bearing capacity of foundation soil are done by settlement analysis and bearing capacity analysis, which were presented in Chaps. 4 and 6, respectively.

The factor of safety against overturning is determined by dividing the total righting moment by the total overturning moment. Since overturning tends to occur about the front base of the wall (at the toe), the righting moments and the overturning moments are computed about the toe of the wall.

The factor of safety against bearing capacity failure is determined by dividing the ultimate bearing capacity by the actual maximum contact (base) pressure. The contact pressure is determined by the methods presented in Chap. 6.

Some common minimum factors of safety for sufficient stability are as follows:

Factor of safety against sliding = 1.5 (if the passive earth pressure of the soil at the toe in front of the wall is neglected) [3][1]

= 2.0 (if the passive earth pressure of the soil at the toe in front of the wall is included) [3][1]

Factor of safety against overturning = 1.5 (granular backfill soil)

= 2.0 (cohesive backfill soil) [1]

Factor of safety against bearing capacity failure = 3.0

The two example problems that follow illustrate the investigation of stability analysis for retaining walls. Example 10-1 refers to a gravity wall, and Example 10-2 refers to a cantilever wall.

EXAMPLE 10-1

Given

1. The retaining wall shown in Fig. 10-4 is to be constructed of concrete having a unit weight of 150 pcf.

2. The retaining wall is to support a deposit of granular soil that has the following properties:

$$\gamma = 115 \text{ pcf}$$

$$\phi = 30°$$

$$c = 0$$

---

[1] Reprinted with permission of Macmillan Publishing Co., Inc., from *Theory and Practice of Foundation Engineering*, by Louis J. Goodman and R. H. Karol. Copyright © 1968, Macmillan Publishing Co., Inc.

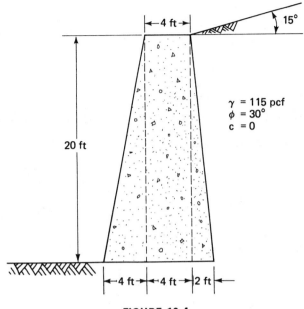

FIGURE 10-4

3. The coefficient of base friction is 0.55.

4. The ultimate bearing capacity of the foundation soil is 6.5 tons/ft².

Required

Check the stability of the proposed retaining wall; that is, check:

1. The factor of safety against overturning.

2. The factor of safety against sliding.

3. The factor of safety against bearing capacity failure.

Solution

**Calculation of the active earth pressure on the back
of the wall by Rankine theory**

From Eqs. (9-1) and (9-2),

$$P_a = \tfrac{1}{2}\gamma H^2 \cos \beta \frac{\cos \beta - \sqrt{\cos^2 \beta - \cos^2 \phi}}{\cos \beta + \sqrt{\cos^2 \beta - \cos^2 \phi}}$$

Referring to Fig. 10-5,

$$H = \overline{BC} = 20 + 2 \tan 15° = 20.54 \text{ ft}$$

$$P_a = (\tfrac{1}{2})(0.115)(20.54)^2(\cos 15°)\,\frac{\cos 15° - \sqrt{\cos^2 15° - \cos^2 30°}}{\cos 15° + \sqrt{\cos^2 15° - \cos^2 30°}}$$
$$= 9.05 \text{ kips/ft}$$

P_a acts parallel to the surface of the backfill, and therefore

Horizontal component $(P_h) = P_a \cos 15° = (9.05) \cos 15° = 8.74$ kips/ft
Vertical component $(P_v) = P_a \sin 15° = (9.05) \sin 15° = 2.34$ kips/ft

Calculation of righting moment (see Fig. 10-5)

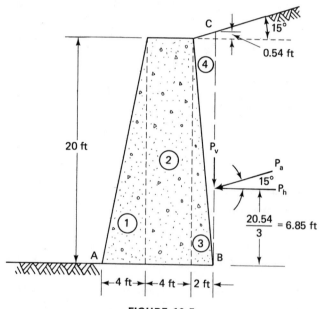

FIGURE 10-5

Component	Weight of Component (kips/ft)		Moment Arm from A (ft)	Righting Moment about A (ft-kips/ft)
1	$(0.15)(\tfrac{1}{2})(4)(20)$	$= 6.00$	$(\tfrac{2}{3})(4) = \tfrac{8}{3}$	16.0
2	$(0.15)(4)(20)$	$= 12.00$	$4 + \tfrac{4}{2} = 6$	72.0
3	$(0.15)(\tfrac{1}{2})(2)(20)$	$= 3.00$	$4 + 4 + (\tfrac{1}{3})(2) = \tfrac{26}{3}$	26.0
4	$(0.115)(\tfrac{1}{2})(20.54)(2)$	$= 2.36$	$4 + 4 + (\tfrac{2}{3})(2) = \tfrac{28}{3}$	22.0
P_v		2.34	$4 + 4 + 2 = 10$	23.4
	$\Sigma V = 25.70$		$\Sigma M_r = 159.4$	

Calculation of overturning moment

Overturning moment $(M_0) = (8.74 \text{ kips/ft})(6.85 \text{ ft}) = 59.9 \text{ ft-kips/ft}$

1. Factor of safety against overturning $= \dfrac{\sum M_r}{\sum M_0} = \dfrac{159.4}{59.9} = 2.66 > 1.5$

    (for granular backfill)   O.K.

2. Factor of safety against sliding $= \dfrac{\text{sliding resistance}}{\text{sliding force}}$

    $= \dfrac{(\text{coefficient of base friction})(\sum V)}{P_h} = \dfrac{(0.55)(25.70)}{8.74} = 1.62 > 1.5$

                                                               O.K.

Base pressure calculations

Location of resultant $R \, (= \sum V)$ if R acts at $\bar{x}$ ft from the toe   (point A)

$$\bar{x} = \frac{\sum M_A}{\sum V} = \frac{\sum M_r - \sum M_0}{\sum V} = \frac{159.4 - 59.9}{25.70} = 3.87 \text{ ft}$$

$$e = \frac{4 + 4 + 2}{2} - 3.87 = 1.13 \text{ ft} < \frac{L}{6} \quad (\text{i.e., } \tfrac{10}{6}, \text{ or } 1.67 \text{ ft}) \quad \text{O.K.}$$

(i.e., R acts within the middle third of the base)

Using the flexural formula, from Eq. (6-4) (see Chap. 6)

$$q = \frac{Q}{A} \pm \frac{M_x x}{I_y} \pm \frac{M_y y}{I_x} \tag{6-4}$$

Here

$$Q = \text{resultant } (R) = \sum V = 25.70 \text{ kips}$$

$$A = (1)(10) = 10 \text{ ft}^2$$

$$M_x = Q \times e = (25.70)(1.13) = 29.04 \text{ ft-kips}$$

$$x = \frac{10}{2} = 5 \text{ ft}$$

$$I_y = \frac{bh^3}{12} = \frac{(1)(10)^3}{12} = 83.33 \text{ ft}^4$$

$$M_y = 0 \quad (\text{one-way bending})$$

$$q = \frac{25.70}{10} \pm \frac{(29.04)(5)}{83.33}$$

$$q_L = 2.57 + 1.74 = 4.31 \text{ kips/ft}^2 = 2.16 \text{ tons/ft}^2$$

$$q_R = 2.57 - 1.74 = 0.83 \text{ kip/ft}^2 = 0.42 \text{ ton/ft}^2$$

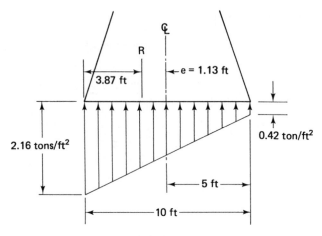

FIGURE 10-6

The pressure distribution is shown in Fig. 10-6.

3. Factor of safety against bearing capacity failure

$$= \frac{\text{ultimate bearing capacity}}{\text{actual maximum base pressure}} = \frac{6.5}{2.16} = 3.01 > 3 \qquad \text{O.K.}$$

EXAMPLE 10-2

Given

1. The retaining wall shown in Fig. 10-7.

2. Backfill material is Type 1 soil (see Fig. 10-3).

3. The unit weight and the ϕ angle of the backfill material are estimated to be 120 pcf and 37°, respectively.

4. The coefficient of base friction is estimated to be 0.45.

5. Allowable soil pressure is 3 kips/ft².

Required

1. The factor of safety against overturning.

2. The factor of safety against sliding. For sliding, analyze both without and with passive earth pressure at toe.

3. The safety against failure of the foundation soil.

Note: Use Fig. 10-3 to compute the lateral earth pressure. Assume that unit weight of concrete is 150 pcf.

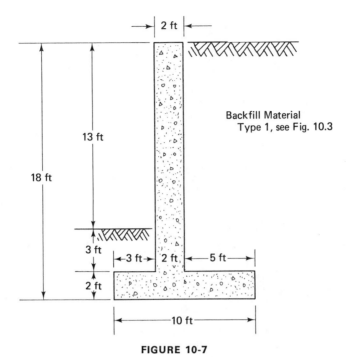

FIGURE 10-7

Solution

Calculation of the active earth pressure by Fig. 10-3

From Fig. 10-3,

$P_h = \frac{1}{2}K_h H^2$ with $\beta = 0°$ and Type 1 backfill material, $K_h = 30$

$P_h = (\frac{1}{2})(30)(18)^2 = 4860$ lb/ft $= 4.86$ kips/ft

$P_v = \frac{1}{2}K_v H^2$ with $\beta = 0°$, $K_v = 0$

$P_v = 0$

Calculation of righting moment (see Fig. 10-8)

Component	Weight of Component (kips/ft)	Moment Arm from Toe (ft)	Righting Moment about Toe (ft-kips/ft)
1	$(0.15)(2)(13 + 3) = 4.8$	$3 + \frac{2}{2} = 4.0$	19.2
2	$(0.15)(2)(10) = 3.0$	$\frac{10}{2} = 5.0$	15.0
3	$(0.12)(5)(13 + 3) = 9.6$	$3 + 2 + \frac{5}{2} = 7.5$	72.0
	$\sum V = 17.4$		$\sum M_r = 106.2$

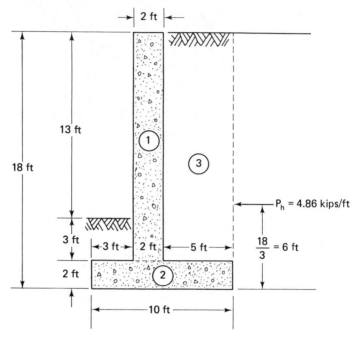

FIGURE 10-8

Calculation of overturning moment

Overturning moment $(M_0) = (4.86 \text{ kips/ft})(6 \text{ ft}) = 29.16 \text{ ft-kips/ft}$

1. Factor of safety against overturning $= \dfrac{\sum M_r}{\sum M_0} = \dfrac{106.2}{29.16} = 3.64 > 1.5$

   O.K.

2. Factor of safety against sliding

   a. Without passive earth pressure analysis (neglect passive earth pressure at the toe)

   Factor of safety against sliding $= \dfrac{\text{sliding resistance}}{\text{sliding force}}$

   $= \dfrac{(\text{coefficient of base friction})(\sum V)}{P_h} = \dfrac{(0.45)(17.4)}{4.86} = 1.61 > 1.5$

   O.K.

   b. With passive earth pressure at the toe

   Sliding resistance = passive earth pressure at toe plus friction available along the base

According to Rankine theory for level backfill, from Eqs. (9-3) and (9-6)

$$P_p = \tfrac{1}{2}\gamma H^2 K_p = \tfrac{1}{2}\gamma H^2 \frac{1 + \sin \phi}{1 - \sin \phi} = (\tfrac{1}{2})(0.12)(5)^2 \frac{1 + \sin 37°}{1 - \sin 37°}$$

$$= 6.03 \text{ kips/ft}$$

$$\text{Factor of safety against sliding} = \frac{(0.45)(17.4) + 6.03}{4.86} = 2.85$$

$$2.85 > 2.0 \quad \text{O.K.}$$

Base pressure calculations

Location of resultant $R \, (= \sum V)$ if R acts at $\bar{x}$ ft from the toe

$$\bar{x} = \frac{\sum M_{\text{toe}}}{\sum V} = \frac{\sum M_r - \sum M_0}{\sum V} = \frac{106.2 - 29.16}{17.4} = 4.43 \text{ ft}$$

$$e = \frac{10}{2} - 4.43 = 0.57 \text{ ft} < \frac{L}{6} \text{ (i.e., } \tfrac{10}{6}, \text{ or } 1.67 \text{ ft)} \quad \text{O.K.}$$

Using the flexural formula, from Eq. (6-4),

$$q = \frac{Q}{A} \pm \frac{M_x x}{I_y} \pm \frac{M_y y}{I_x} \tag{6-4}$$

Here

$$Q = \text{resultant } (R) = \sum V = 17.4 \text{ kips}$$

$$A = (1)(10) = 10 \text{ ft}^2$$

$$M_x = Q \times e = (17.4)(0.57) = 9.92 \text{ ft-kips}$$

$$x = \frac{10}{2} = 5 \text{ ft}$$

$$I_y = \frac{bh^3}{12} = \frac{(1)(10)^3}{12} = 83.33 \text{ ft}^4$$

$$M_y = 0$$

$$q = \frac{17.4}{10} \pm \frac{(9.92)(5)}{83.33}$$

$$q_L = 1.74 + 0.60 = 2.34 \text{ kips/ft}^2$$

$$q_R = 1.74 - 0.60 = 1.14 \text{ kips/ft}^2$$

3. Since $q_L = 2.34$ kips/ft², which is less than the allowable soil pressure of 3.0 kips/ft² (given), the wall is safe against failure of the foundation soil.

10-4 BACKFILL DRAINAGE

If water is allowed to permeate the soil behind a retaining wall, large additional pressure will be applied to the wall. Unless the wall is designed to withstand this large additional pressure (not the usual practice), it is imperative that steps be taken to prevent water that infiltrates the backfill soil from accumulating behind the wall.

One method of preventing water from accumulating behind a wall is to provide an effective means of draining away any water that enters the backfill soil. To accomplish this, it is highly desirable to use as backfill material a highly pervious soil such as sand, gravel, or crushed stone. To remove water from behind the wall, 4- to 6-in. weep holes, which are pipes extending through the wall (see Fig. 10-9a), may be placed every 5 to 10 ft along the wall. A perforated drain pipe placed longitudinally along the back of the wall (Fig. 10-9b) may also be used to remove water from behind the wall. In this case, the pipe is surrounded by filter material and water drains through the filter material into the pipe and then through the pipe to one end of the wall. In both cases (weep holes and drain pipes) a filter material must be placed adjacent to the pipe to prevent clogging, and the pipes must be kept clear of debris.

If a less pervious material (silt, granular soil containing clay, etc.) has to be used as backfill because a free-draining, granular material is expensive in the locality, it is highly desirable to place a wedge of pervious material adjacent to the wall, as shown in Fig. 10-10. If this is not possible, a "drainage blanket" of pervious material may be placed as shown in Fig. 10-11.

A highly impervious soil (clay) is very undesirable as backfill material because, in addition to the excessive lateral earth pressure it will create, it also is difficult to drain and may be subject to frost action. Also, clays are subject to swelling and shrinking. If clay soil must be used as backfill material, it would be advisable to place a wedge of pervious material adjacent to the wall between the wall and the clay backfill (as shown in Fig. 10-10) [6].

10-5 SETTLEMENT AND TILTING

A certain amount of settlement by retaining walls is to be expected, just as by any other structures resting on footings or piles. In the case of retaining walls on granular soils, most of the expected settlement will occur by the time the construction of the wall and placement of backfill have been completed. In the case of retaining walls on cohesive soils, for which consolidation theory is applicable, settlement will occur slowly and for a long period of time after construction has been completed.

The amount of settlement for retaining walls resting on spread footings

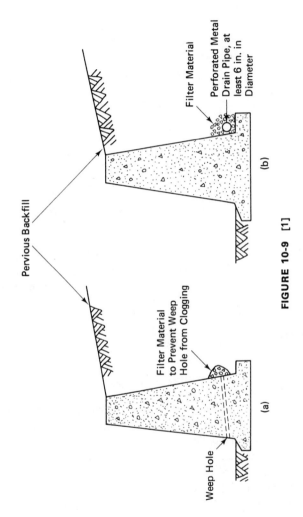

FIGURE 10-9 [1]

269

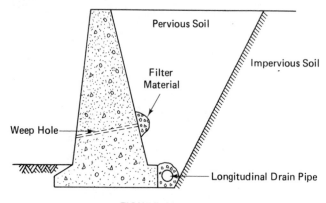

FIGURE 10-10 [4]

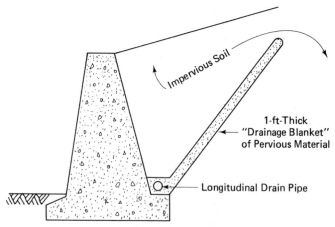

FIGURE 10-11 [5][1]

can be determined using the principles of settlement analysis for footings (see Chap. 4). For walls resting on piles, the amount of settlement can be determined using the principles of settlement analysis for pile foundations (see Chap. 7). To keep settlement relatively uniform, the resultant force must be kept near the middle of the base.

If the soil upon which a retaining wall rests is not uniform in bearing capacity along the length of the wall, differential settlement may occur along the wall, which could cause the wall to crack vertically. If soil of poor bearing capacity occurs only for a very short distance, differential settlement may not be a problem, as the wall tends to bridge across poor material. If, however, the poor bearing capacity of the soil exists for a considerable distance along the length of the wall, differential settlement will likely happen unless the

---

[1] Reprinted with permission of Macmillan Publishing Co., Inc. from *Introductory Soil Mechanics and Foundations*, 4th Edition, by George F. Sowers. Copyright © 1979, Macmillan Publishing Co., Inc.

designer takes this into account and implements remedies to correct the situation found in this particular case. Possible remedies include improving the soil (e.g., by replacement, compaction, or stabilization of the soil) and changing the width of the footing. If computed settlement is excessive, pile foundations may be used [7].

In addition to settlement, a retaining wall is also subject to tilting caused by eccentric pressure on the base of the wall. Tilting can be reduced by keeping the resultant force near the middle of the base. In many cases, the wall tilts forward because the resultant force intersects the base at a point between the center and the toe [1].

It is difficult to determine the amount of tilting to be expected, and rough estimates must suffice. If the stability requirements are met in accordance with established design procedures (see Sec. 10-3), the amount of tilting may be expected to be in the order of magnitude one-tenth of 1 % of the height of wall or less. However, if the subsoil consists of a compressive layer, this amount may be exceeded [1].

10-6 PROBLEMS

10-1 A proposed L-shaped reinforced concrete retaining wall is shown in Fig. 10-12. The backfill material will be Type 2 soil (Fig. 10-3) and its unit weight is

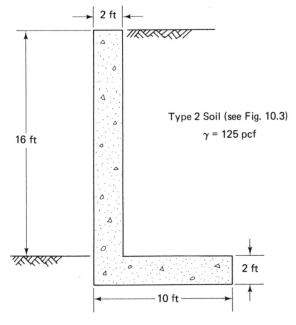

FIGURE 10-12

125 pcf. The coefficient of base friction is estimated to be 0.48. The allowable soil pressure for the foundation soil is 4 kips/ft². Determine (1) the factor of safety against overturning, (2) the factor of safety against sliding, and (3) the safety against failure of the foundation.

10-2 Investigate the stability against overturning, the sliding resistance (consider passive earth pressure at the toe), and the foundation soil pressure of the retaining wall shown in Fig. 10-13. The retaining wall is to support a deposit of granular soil, which has a unit weight of 110 pcf and an angle of internal friction of 32°. The coefficient of base friction is estimated to be 0.50. The allowable soil pressure for the foundation soil is 3 kips/ft². *Note:* Use Rankine theory to calculate both active and passive earth pressures.

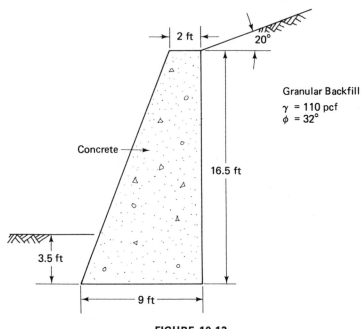

FIGURE 10-13

10-3 For the retaining wall shown in Fig. 10-14, compute the factor of safety against overturning and the factor of safety against sliding (analyze the latter both without and with passive earth pressure at toe). Also determine the soil pressure at the base of the wall. Use the Rankine equation to compute passive earth pressure.

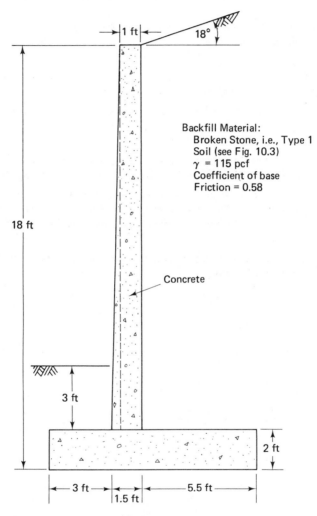

Backfill Material:
Broken Stone, i.e., Type 1
Soil (see Fig. 10.3)
γ = 115 pcf
Coefficient of base
Friction = 0.58

Concrete

FIGURE 10-14

References

[1] WAYNE C. TENG, *Foundation Design*, Prentice-Hall, Inc., Englewood Cliffs, N.J., 1962.

[2] AREA, *Manual of Recommended Practice*, Construction and Maintenance Section, Engineering Division, Association of American Railroads, Chicago, 1958.

[3] LOUIS J. GOODMAN AND R. H. KAROL, *Theory and Practice of Foundation Engineering*, Macmillan Publishing Co., Inc., New York, 1968.

[4] B. K. HOUGH, *Basic Soils Engineering*, 2nd ed., The Ronald Press Company, New York, 1969. Copyright © 1969, by John Wiley & Sons, Inc. Reprinted by permission of John Wiley & Sons, Inc.

[5] GEORGE F. SOWERS, *Introductory Soil Mechanics and Foundations: Geotechnical Engineering*, 4th ed., Macmillan Publishing Co., Inc., New York, 1979.

[6] RALPH B. PECK, WALTER E. HANSEN, AND THOMAS H. THORNBURN, *Foundation Engineering*, 2nd ed., John Wiley & Sons, Inc., New York, 1974. Copyright © 1974, by John Wiley & Sons, Inc. Reprinted by permission of John Wiley & Sons, Inc.

[7] JOSEPH E. BOWLES, *Foundation Analysis and Design*, McGraw-Hill Book Company, New York, 1968.

11

Soil Compaction

11-1 DEFINITION AND PURPOSE OF COMPACTION

The general meaning of the verb "compact" is to press closely together. In the field of soil mechanics, it means to press the soil particles tightly together by expelling air from the void space. Compaction is normally produced deliberately and proceeds rapidly during construction, often by heavy compaction rollers. This is in contrast to "consolidation" (Chap. 4), which also results in a reduction of voids but which is caused by extrusion of water (rather than air) from the void space. Also, consolidation is not rapid.

Compaction of soil increases its density and produces three important effects: (1) an increase in shear strength of the soil, (2) a decrease in future settlement of the soil, and (3) a decrease in its permeability [1]. These three effects are beneficial for various types of earth construction, such as highways, airfields, and earth dams; and, as a general rule, the greater the compaction, the greater these benefits will be. Compaction is actually a rather cheap and effective way to improve the properties of a given soil.

The amount of compaction is quantified in terms of the density (dry unit weight) of the soil. [The dry unit weight, γ_d, can be computed in terms of the wet unit weight, γ_{wet}, and the moisture content, w (expressed as a decimal), by Eq. (11-1).]

$$\gamma_d = \frac{\gamma_{wet}}{1 + w} \qquad (11\text{-}1)$$

As a general rule, dry soils can be best compacted (and thus a greater density achieved) if for each soil a specified amount of water is added to it. In effect, the water acts as a lubricant and allows the soil particles to be packed together better. If, however, too much water is added, a lesser density would result. Thus, for a given compactive effort, there is a particular moisture content at which the dry density is greatest and the compaction is best. This moisture content is called the "optimum moisture content," and the associated dry density is called the "maximum dry density."

The usual practice in a construction project is to perform laboratory compaction tests (covered in Sec. 11-2) on representative samples from the construction site to determine the optimum moisture content and the maximum dry density. This maximum dry density is used by the designer in specifying the design shear strength, resistance to future settlement, and permeability characteristics. The soil is then compacted by field compaction methods (covered in Sec. 11-3) to achieve the laboratory maximum dry density (or a percentage of it). In-place soil density tests (covered in Sec. 11-4) are used to determine if the laboratory maximum dry density (or an acceptable percentage thereof) has been achieved. Section 11-5 covers field control of compaction.

11-2 LABORATORY COMPACTION TESTS

As related in the last section, laboratory compaction tests are performed to determine the optimum moisture content and the maximum dry density. Compaction test equipment, shown in Fig. 11-1, consists of a base plate, collar, and mold, in which the soil is placed, and of a hammer that is raised and dropped freely onto the soil. The size of the mold and both the weight of the hammer and its drop distance are standardized, with several variations in size and weight available (see Table 11-1).

To carry out a laboratory compaction test (known as a *Proctor Test*), the soil sample from the field is allowed to dry until it becomes friable under a trowel. The soil sample may be dried in the air or in a drying oven. If an oven is used, the oven temperature should not exceed 60°C (140°F). After drying, a representative sample is taken and thoroughly mixed with water to obtain a low water content. This prepared sample is then placed in a compaction mold (with collar attached) and compacted in layers by dropping the hammer onto the soil sample in the mold a specified distance and a specified number of uniformly distributed blows per layer. This results in a specified energy exertion per unit volume of soil. Upon completion of the compaction, the attached collar is removed, and the compacted soil is trimmed until it is even with the top of the mold. The wet density of the compacted soil specimen is then determined by dividing the weight of compacted soil in the mold by the volume of the soil specimen, which is the volume of the mold.

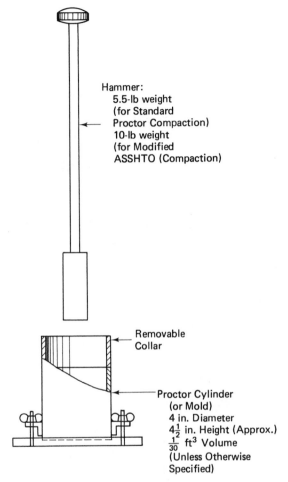

Hammer:
5.5-lb weight
(for Standard
Proctor Compaction)
10-lb weight
(for Modified
ASSHTO (Compaction)

Removable
Collar

Proctor Cylinder
(or Mold)
4 in. Diameter
$4\frac{1}{2}$ in. Height (Approx.)
$\frac{1}{30}$ ft^3 Volume
(Unless Otherwise
Specified)

FIGURE 11-1 Compaction test equipment. [2]

The compacted soil is subsequently removed from the mold, and its moisture content is then determined. With the wet density of the compacted soil and the moisture content known, the dry density is computed using Eq. (11-1).

The next step is to add more water to, and mix the water with, the remaining portion of the soil sample being tested and to repeat the compaction procedure (described in the previous paragraph). Again, the wet density and moisture content are determined, with the wet density expected to be greater than that of the previous (drier) specimen. The dry density of this specimen can also be computed using Eq. (11-1).

Additional water is added to the remaining portion of the sample and the entire procedure is repeated. Eventually, a point will be reached where addi-

TABLE 11-1 Summary of specifications for compaction testing equipment, compaction effort, and sample fractionation [2].

TEST DESIGNATION

	AASHTO: T 99-61 ASTM D 698-64T				AASHTO: T 180-61 ASTM D 1557-64T			
	Method A[1]	Method B	Method C[2]	Method D	Method A[3]	Method B	Method C[2]	Method D
Hammer weight (lb)	5.5	5.5	5.5	5.5	10	10	10	10
Drop (in.)	12	12	12	12	18	18	18	18
Size of mold								
Diameter (in.)	4	6	4	6	4	6	4	6
Height (in.)	4.58	4.58	4.58	4.58	4.58	4.58	4.58	4.58
Volume (ft³)	$\frac{1}{30}$	$\frac{1}{13.33}$	$\frac{1}{30}$	$\frac{1}{13.33}$	$\frac{1}{30}$	$\frac{1}{13.33}$	$\frac{1}{30}$	$\frac{1}{13.33}$
Number of layers	3	3	3	3	5	5	5	5
Blows per layers	25	56	25	56	25	56	25	56
Fraction tested	$-$No. 4	$-$No. 4	$-\frac{3}{4}$ in.	$-\frac{3}{4}$ in.	$-$No. 4	$-$No. 4	$-\frac{3}{4}$ in.	$-\frac{3}{4}$ in.

[1]This is the original "Standard Proctor" test.
[2]Note 2 of this method provides for "stone substitution," a means of evaluating the effect of stone sizes up to 2-in. diameter. (See complete specification.)
[3]This is the original "modified AASHTO" test.

tional water causes a decrease or no change in the wet density of the sample. (For detailed laboratory procedures, the reader is referred to a soil-testing manual.)

A plot is made of moisture content versus soil dry density for the data collected as described above, and the result will be of a form similar to the curve shown in Fig. 11-2. The coordinates of the point at the peak of the

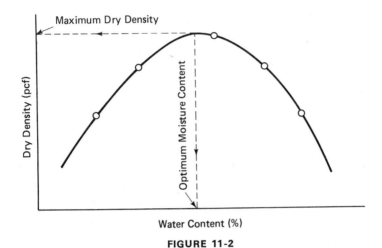

FIGURE 11-2

curve give the maximum dry density and the optimum moisture content. Presumably, this gives the maximum expected density—the density to be used by the designer and the density to be striven for in the field compaction. To achieve this maximum density, the field compaction should be done at or near the optimum moisture content.

Example 11-1 illustrates the computation of the density of a specimen of a laboratory compacted soil. Example 11-2 illustrates the determination of maximum dry density and optimum moisture content, as the result of a laboratory compaction test.

EXAMPLE 11-1

Given

1. The combined weight of a mold and the specimen of compacted soil it contains is 8.63 lb.

2. The volume of the mold is $\frac{1}{30}$ ft^3.

3. The weight of the mold is 4.35 lb.

4. The water content of the specimen is 10%.

Required

1. The wet density of the specimen.

2. The dry density of the specimen.

Solution

1. Wet density of the specimen (γ_{wet}):

$$\gamma_{wet} = \frac{\text{wet weight of specimen}}{\text{volume of specimen}}$$

$$= \frac{8.63 - 4.35}{\frac{1}{30}} = 128.4 \text{ pcf}$$

2. Dry density of the specimen (γ_d):

$$\gamma_d = \frac{\gamma_{wet}}{1 + w} \tag{11-1}$$

$$\gamma_d = \frac{128.4}{1 + 0.1} = 116.7 \text{ pcf}$$

EXAMPLE 11-2

Given

A set of laboratory compaction test data and results is tabulated below. The test was conducted in accordance with ASTM D 698 Standard Proctor Test.

Determination Number	1	2	3	4	5
Dry Density (pcf)	112.2	116.7	118.3	115.2	109.0
Moisture Content (%)	7.1	10.0	13.4	16.7	20.1

Required

1. Plot a Proctor curve (i.e., dry density versus water content on graph paper).

2. Determine the maximum dry density and optimum moisture content.

Solution

1. See Fig. 11-3.

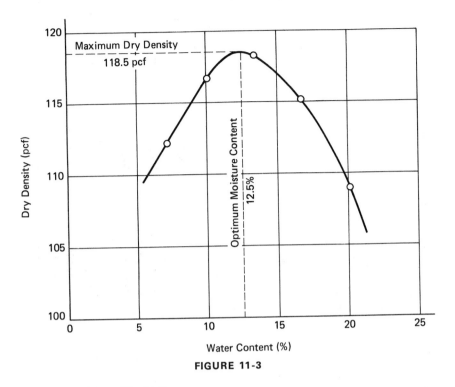

FIGURE 11-3

2. From Fig. 11-3

$$\text{Maximum dry density} = 118.5 \text{ pcf}$$
$$\text{Optimum moisture content} = 12.5\%$$

The type of soil is the primary factor affecting the maximum dry density and optimum moisture content for a given compactive effort and compaction method. Maximum dry densities may range from about 60 pcf for organic soils to about 145 pcf for well-graded granular material containing just enough fines to fill small voids. Optimum moisture content may range from around 5% for granular material to about 35% for elastic silts and clays. Higher optimum moisture contents are generally associated with lower dry densities. Higher dry densities are associated with well-graded granular materials. Uniformly graded sand, clays of high plasticity, and organic silts and clays typically respond poorly to compaction [3].

The following two tables give some general compaction characteristics of various types of soil. Table 11-2 gives such characteristics along with the values as embankment, subgrade, and base material for soils classified according to the "Unified soil classification system." Table 11-3 gives anticipated embankment performance for soils classified according to the AASHTO system.

TABLE 11-2 Compaction characteristics and ratings of Unified soil classification classes for soil construction [3, 4].

Class	Compaction Characteristics	Maximum Dry Density Standard AASHTO (pcf)	Compressibility and Expansion	Value as Embankment Material	Value as Subgrade Material	Value as Base Course
GW	Good: tractor, rubber-tired, steel wheel, or vibratory roller	125–135	Almost none	Very stable	Excellent	Good
GP	Good: tractor, rubber-tired, steel wheel, or vibratory roller	115–125	Almost none	Reasonably stable	Excellent to good	Poor to fair
GM	Good: rubber-tired or light sheepsfoot roller	120–135	Slight	Reasonably stable	Excellent to good	Fair to poor
GC	Good to fair: rubber-tired or sheepsfoot roller	115–130	Slight	Reasonably stable	Good	Good to fair
SW	Good: tractor, rubber-tired, or vibratory roller	110–130	Almost none	Very stable	Good	Fair to poor
SP	Good: tractor, rubber-tired, or vibratory roller	100–120	Almost none	Reasonably stable when dense	Good to fair	Poor
SM	Good: rubber-tired or sheepsfoot roller	110–125	Slight	Reasonably stable when dense	Good to fair	Poor
SC	Good to fair: rubber-tired or sheepsfoot roller	105–125	Slight to medium	Reasonably stable	Good to fair	Fair to poor
ML	Good to poor: rubber-tired or sheepsfoot roller	95–120	Slight to medium	Poor stability, high density required	Fair to poor	Not suitable
CL	Good to fair: sheepsfoot or rubber-tired roller	95–120	Medium	Good stability	Fair to poor	Not suitable

TABLE 11-2 Continued

Class	Compaction Characteristics	Maximum Dry Density Standard AASHTO (pcf)	Compressibility and Expansion	Value as Embankment Material	Value as Subgrade Material	Value as Base Course
OL	Fair to poor: sheepsfoot or rubber-tired roller	80–100	Medium to high	Unstable, should not be used	Poor	Not suitable
MH	Fair to poor: sheepsfoot or rubber-tired roller	70–95	High	Poor stability, should not be used	Poor	Not suitable
CH	Fair to poor: sheepsfoot roller	80–105	Very high	Fair stability, may soften on expansion	Poor to very poor	Not suitable
OH	Fair to poor: sheepsfoot roller	65–100	High	Unstable, should not be used	Very poor	Not suitable
PT	Not suitable	—	Very high	Should not be used	Not suitable	Not suitable

TABLE 11-3 General guide to selection of soils on basis of anticipated embankment performance [3, 5].

HRB Classification	Visual Description	Maximum Dry-Weight Range (pcf)	Optimum Moisture Range (%)	Anticipated Embankment Performance
A-1-a A-1-b	Granular material	115–142	7–15	Good to excellent
A-2-4 A-2-5 A-2-6 A-2-7	Granular material with soil	110–135	9–18	Fair to excellent
A-3	Fine sand and sand	110–115	9–15	Fair to good
A-4	Sandy silts and silts	95–130	10–20	Poor to good
A-5	Elastic silts and clays	85–100	20–35	Unsatisfactory
A-6	Silt-clay	95–120	10–30	Poor to good
A-7-5	Elastic silty clay	85–100	20–35	Unsatisfactory
A-7-6	Clay	90–115	15–30	Poor to fair

11-3 FIELD COMPACTION

Normally, soil is compacted in layers. An approximately 8-in. loose horizontal layer of soil is often spread from trucks and then compacted to a thickness of about 6 in. The moisture content can be increased by sprinkling water over the soil if the soil is too dry and thoroughly mixing it into the uncompacted soil by disk plowing. If the soil is too wet, the moisture content can be reduced by aeration (i.e., by spreading the soil in the sun and turning it with a disk plow to provide aeration and drying.) The actual compaction is done by *tampers* and/or *rollers* and is normally accomplished with a maximum of 6 to 10 complete coverages of the compaction equipment. The surface of each compacted layer should be scarified by disk plowing or other means to provide bonding between layers. Various kinds of field compaction equipment (i.e., tampers and rollers) will be discussed briefly in this section.

A tamper is a device that compacts by delivering a succession of relatively light, vertical blows. Tampers are held in place and operated by hand. They may be powered either pneumatically or by gasoline-driven pistons. Tampers are limited in scope and compacting ability. Therefore, tampers are useful in areas not readily accessible to rollers, in which case the soil may be placed in loose horizontal layers not exceeding 6 in. and then compacted with tampers.

Rollers come in a variety of forms, such as the smooth wheel roller, the sheepsfoot roller, the pneumatic roller, and the vibrating roller. Some of these are self-propelled, while some are towed by tractors. Some of these are more suited to certain types of soil. Rollers can easily cover large areas relatively quickly and with great compacting pressures. A brief description of the four types of roller mentioned above follows.

The *smooth wheel roller* (Fig. 11-4) employs two or three smooth metal rollers. It is useful in compacting base courses and paving mixtures and is

FIGURE 11-4   Smooth wheel roller. (Photo courtesy of Koehring Company)

also used to provide a smooth finished grade. Generally, smooth wheel rollers are self-propelled and are equipped with a reversing gear so that they can be driven back and forth without turning. The smooth wheel roller provides compactive effort primarily through its static weight.

The *sheepsfoot roller* (Fig. 11-5) consists of a drum with metal projecting "feet" attached. Since only the projecting feet come in contact with the soil, the area of contact between roller and soil is smaller (than for a smooth metal roller) and therefore a greater compacting pressure results (generally more than 200 psi). The sheepsfoot roller provides kneading action and is effective for compacting fine-grained soils (such as clay and silt).

The *pneumatic roller* (Fig. 11-6) consists of a number of rubber tires, highly inflated. These vary from small ones to very large and heavy ones. Most large pneumatic rollers are towed while some smaller ones are self-propelled. Some have boxes mounted above the wheels, to which sand or other material can be added for increased compacting pressure. Clayey soils

FIGURE 11-5 Sheepsfoot roller. (Photo courtesy of Koehring Company)

FIGURE 11-6 Pneumatic roller. (Photo courtesy of Koehring Company)

and silty soils may be compacted effectively by pneumatic rollers. They are also effective in compacting granular material containing some small amount of fines.

The *vibratory roller* contains some kind of vibrating unit that imparts an up and down vibration to the roller as the roller is pulled over the soil. Vibrating units can supply frequencies of vibration at 1500 to 2000 cycles per minute, depending on compaction requirements. These are effective in compacting granular materials, particularly clean sand and gravels.

Two means (or possibly a combination of the two) may be used to specify the compaction requirement. One is to specify the procedure to be followed by the contractor, such as the type of compactor (i.e., roller) to be used and the number of passes to be made. The other is to simply specify the required final dry density of the compacted soil. The first method has the advantage that little testing is required, but it has the disadvantage that the specified procedure may not produce the required result. The second method requires much field testing, but it ensures that the required dry density is achieved. In effect, the second method specifies the required final dry density but leaves it up to the contractor as to how that density is achieved. This (i.e., the second) method is probably more commonly used.

11-4 IN-PLACE SOIL DENSITY TEST

As related previously, after a fill layer of soil has been compacted by the contractor, it is important that the in-place dry density of the compacted soil be determined in order to ascertain whether the maximum laboratory dry density has been attained. If the maximum density (or an acceptable percentage thereof) has not been attained, additional compaction effort is required.

There are several methods for determining in-place density. As a general rule, the weight and the volume of an in-place soil sample are determined, from which the density can be computed. The measurement of the weight of the sample is straightforward, but there are several methods for determining the volume of the sample. In the case of cohesive soils, a thin-walled cylinder may be driven into the soil to remove a sample. The volume of the soil sample is known from the volume of the cylinder. This method is known as "density of soil in-place by the drive cylinder method" and is designated as ASTM D 2937 or AASHTO T 204. This "drive cylinder method" is not applicable for very hard soil that cannot be easily penetrated. Neither is it applicable for low plasticity or cohesionless soils, which are not readily retained in the cylinder.

For low plasticity or cohesionless soils, a hole can be dug and the volume determined by filling it with loose, dry sand of uniform density (such as Ottawa sand). An alternative method is to fill the hole with water (utilizing a rubber membrane, or balloon). In either case, the volume of the hole, and thus the volume of the sample removed, is measured by the volume of material (sand or water) added. The volume of the hole determined by filling it with sand is called "density of soil in-place by the sand-cone method" and is designated as ASTM D 1556 or AASHTO T 191. The volume of hole determined by filling it with water is called "density of soil in-place by the rubber-balloon method" and is designated as ASTM D 2167 or AASHTO T 205.

In-place density of soil can also be determined through the use of nuclear equipment, which utilizes radioactive materials. This method is called "density of soil and soil-aggregate in-place by nuclear methods" and is designated as ASTM D 2922 or AASHTO T 238.

For detailed instructions regarding all these tests, the reader is referred to ASTM or AASHTO manuals.

In addition to determining in-place wet density of soil, it is also necessary to determine the moisture content of the soil in order to compute the dry density of the compacted soil. Although moisture content can be determined by oven drying, this method is often too time-consuming, since test results are commonly needed quickly. The drying of the soil sample can be accomplished by placing the soil sample in a skillet and placing it over an open flame of a camp stove. The Speedy Moisture Tester (Fig. 11-7) can also be used to determine the moisture content quickly with fairly good results. Because of the rather small amount of sample utilized in this test, the Speedy Moisture Tester may not be appropriate for use in coarser materials.

FIGURE 11-7   Speedy moisture tester. (Photo courtesy of Soiltest, Inc.)

EXAMPLE 11-3

Given

During the construction of a soil embankment, a sample is taken from the compacted embankment for a sand-cone in-place density test. The following data are obtained:

1. Weight of sand used to fill test hole and funnel of sand-cone device = 867 g.
2. Weight of sand to fill funnel = 319 g.
3. Density of sand = 98 lb/ft³.
4. Weight of wet soil from the test hole = 747 g.
5. Moisture content of soil from test hole determined by Speedy Moisture Tester = 13.7%.

Required

Dry unit weight of the compacted soil.

Solution

The weight of sand used in test hole
   = weight of sand used to fill test hole and funnel − weight of sand to fill funnel

= 867 − 319 = 548 g

$$\text{Volume of test hole} = \frac{548/453.6}{98} = 0.0123 \text{ ft}^3$$

$$\text{Wet density of soil in-place} = \frac{747/453.6}{0.0123} = 133.9 \text{ pcf}$$

Dry density of compacted soil (γ_d):

$$\gamma_d = \frac{\gamma_{\text{wet}}}{1 + w} \tag{11-1}$$

$$\gamma_d = \frac{133.9}{1 + 0.137} = 117.8 \text{ pcf}$$

11-5 FIELD CONTROL OF COMPACTION

As related previously, after a fill layer of soil has been compacted, an in-place soil density test is usually performed to determine whether the maximum laboratory dry density (or an acceptable percentage thereof) has been attained. It is common to specify a required percent of compaction, which is "the in-place dry density" divided by "the maximum laboratory dry density" expressed as a percentage, in a contract document. Thus, if the maximum dry density obtained from ASTM or AASHTO compaction in the laboratory is 100 pcf and the required percent of compaction is 95% according to a contract, an in-place density of 95 pcf (or higher) would be acceptable. In theory, this is simple enough to do; but there are some practical considerations that must be taken into account. For example, the type of soil or compaction characteristics of soil taken from borrow pits may vary from one location to another. Also, the degree of compaction may not be uniform throughout.

To deal with the problem of nonuniformity of soil from borrow pits, it is necessary to conduct ASTM or AASHTO compaction tests in the laboratory to establish the maximum laboratory dry density along with the optimum moisture content for each type of soil encountered in the project. Then, as soil is transported from the borrow pit and subsequently placed and compacted in the fill area, it is imperative that the results of each in-place soil density test (i.e., in-place dry density) be checked against the maximum laboratory dry density of the respective type of soil.

To deal with the problem of variable degree of field compaction of soil, it is common practice to specify a minimum number of field density tests. For example, for a dam embankment, it might be specified that one test be made for every 2400 yd³ (loose measure) of fill placed.

To ensure that the required field density is achieved by the field compaction, a specifications contract between the owner and the contractor is prepared. The contract will normally specify the required percent of compaction and the minimum number of field density tests required. For compaction adjacent to a structure, where settlement is a serious matter, a higher percent of compaction and a higher minimum number of tests may be specified than for compaction, for example, of the foundation of a parking lot. The specifications contract may also include additional items, such as maximum thickness of loose lifts (layers) prior to compaction, methods to obtain maximum dry density (e.g., ASTM D 698 or AASHTO T 99), methods to determine in-place density (e.g., ASTM D 1556 or AASHTO T 191), and so on.

As the owner's representative, the soils engineer or technologist is responsible for seeing that the provisions of the contract are carried out precisely and completely. He or she is responsible for the testing and must see that the required compacted dry density is achieved. If a particular test indicates that the required compacted dry density has not been achieved, he or she must require additional compaction effort, possibly including an adjustment in moisture content. In addition, he or she must be knowledgeable and capable of dealing with field situations that arise that may go beyond the "textbook procedure."

EXAMPLE 11-4

Given

1. The soil from a borrow pit to be used for the construction of an embankment gave the following laboratory results when subjected to ASTM D 698 Standard Proctor test (see Example 11-2).

Maximum dry density $= 118.5$ pcf

Optimum moisture content $= 12.5\%$

2. The contractor, during the construction of soil embankment, achieved the following (see Example 11-3):

Dry density reached by field compaction $= 117.8$ pcf

Actual water content $= 13.7\%$

Required

Determine the percent of compaction achieved by the contractor.

Solution

Percent of Standard Proctor compaction achieved

$$= \frac{\text{In-place dry density}}{\text{Maximum laboratory dry density}} \times 100 = \frac{117.8}{118.5} \times 100 = 99.4\%$$

11-6 PROBLEMS

11-1 A compaction test was conducted in a soils laboratory and the Standard Proctor Compaction Procedure (ASTM D 698) was used. The weight of a compacted soil specimen plus weight of mold was determined to be 3815 grams. The volume and weight of mold were $1/30$ ft^3 and 2050 grams, respectively. The water content of the specimen was 9.1%. Compute both wet and dry densities of the compacted specimen.

11-2 A soil sample was taken from the site of a proposed borrow pit and sent to the laboratory for a Standard Proctor Test (ASTM D 698). The results of the test are listed as follows:

Determination Number	1	2	3	4	5
Dry Density (pcf)	107.0	109.8	112.0	111.6	107.3
Moisture Content (%)	9.1	11.8	14.0	16.5	18.9

Plot a moisture content versus dry density curve and determine the maximum dry density and optimum moisture content.

11-3 During the construction of a highway project, a soil sample was taken from the compacted earth fill for a sand-cone in-place density test. The following data were obtained during the test:

1. Weight of sand used to fill test hole and funnel of sand-cone device $= 845$ g.
2. Weight of sand to fill funnel $= 323$ g.

3. Density of sand $= 100$ pcf.
4. Weight of wet soil from test hole $= 648$ g.
5. Moisture content of soil from test hole $= 16\%$.

Calculate dry unit weight (or dry density) of the compacted earth fill.

11-4 A soil sample was taken from a proposed cut area in a highway construction project and was sent to a soils laboratory for a compaction test, using the Standard Proctor Compaction Procedure. The results of the test are as follows:

Maximum dry density $= 112.6$ pcf

Optimum moisture content $= 15.5\%$

The contractor, during the construction of the soil embankment, achieved the following:

Dry density reached by field compaction $= 107.1$ pcf

Actual water content $= 16.0\%$

Determine the percent compaction achieved by the contractor.

References

[1] T. William Lambe, *Soil Testing for Engineers*, John Wiley & Sons, Inc., New York, 1951. Copyright © 1951, by John Wiley & Sons, Inc. Reprinted by permission of John Wiley & Sons, Inc.

[2] B. K. Hough, *Basic Soils Engineering*, 2nd ed., The Ronald Press Company, New York, 1969. Copyright © 1969, by John Wiley & Sons, Inc. Reprinted by permission of John Wiley & Sons, Inc.

[3] Robert D. Krebs and Richard D. Walker, *Highway Materials*, McGraw-Hill Book Company, New York, 1971.

[4] U.S. Army Corps of Engineers, *The Unified Soil Classification System*, Waterways Exp. Sta. Tech. Mem. 3-357 (including Appendix A, 1953, and Appendix B, 1957), Vicksburg, Miss., 1953.

[5] L. E. Gregg, "Earthwork," in K. B. Woods, ed., *Highway Engineering Handbook*, McGraw-Hill Book Company, New York, 1960.

12

Stability Analysis of Slopes

12-1 INTRODUCTION

Whenever a mass of soil has an inclined surface, the potential always exists for part of the soil mass to slide from a higher location to a lower location. Sliding will occur if shear stresses developed in the soil exceed the corresponding shear strength of the soil. This phenomenon is of importance in the case of highway cuts and fills, embankments, earth dams, and so on.

The principle stated above—that sliding will occur if shear stresses developed in the soil exceed the corresponding shear strength the soil possesses—is simple in theory; but certain practical considerations make precise stability analyses of slopes difficult in practice. In the first place, sliding may occur along a number of possible surfaces. In the second place, the shear strength of a given soil generally varies throughout time, as soil moisture and other factors change. Obviously, stability analysis should be done using the smallest shear strength the soil will ever have in the future. This is difficult, if not impossible, to ascertain. It is normal in practice to use appropriate safety factors when making slope stability analyses.

There are several techniques available for stability analysis. Section 12-2 covers the analysis of a soil mass resting on an inclined layer of impermeable soil. Section 12-3 covers two methods of analyzing stability for homogeneous soil. The first is known as the *Culmann method.* It is only applicable to vertical, or nearly vertical, slopes. The second might be called the *stability number method.* Section 12-4 covers the *method of slices.*

12-2 ANALYSIS OF MASS RESTING ON INCLINED LAYER OF IMPERMEABLE SOIL

One situation of slope stability analysis that is fairly simple to analyze is that of a mass resting on an inclined layer of impermeable soil (see Fig. 12-1). There exists a tendency for the upper mass to slide downward along the plane of contact between the upper mass and the lower layer of impermeable soil.

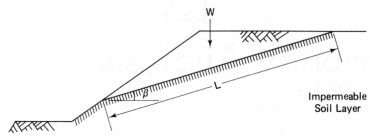

FIGURE 12-1

The force tending to cause sliding is the component along the plane of contact of the weight of the upper mass of soil. The forces tending to resist sliding result from cohesion and friction. In quantitative terms, the cohesion component is obtained by multiplying the unit cohesion by the length L in Fig. 12-1. The friction component is obtained by multiplying the coefficient of friction between the upper mass and the lower layer of impermeable soil by the component perpendicular to the plane of contact of the weight of the upper mass of soil. The factor of safety against sliding is determined by dividing the sum of the forces tending to resist sliding by the force tending to cause sliding. Table 12-1 gives the significance of such factors of safety for design.

TABLE 12-1   Significance of factors of safety for design [1].[1]

Safety Factor	Significance
Less than 1.0	Unsafe
1.0–1.2	Questionable safety
1.3–1.4	Satisfactory for cuts, fills; questionable for dams
1.5–1.75	Safe for dams

[1]Reprinted with permission of Macmillan Publishing Co., Inc., from *Introductory Soil Mechanics and Foundations*, 4th Edition, by George F. Sowers. Copyright © 1979, Macmillan Publishing Co., Inc.

Example 12-1 illustrates the computation of the factor of safety for the analysis of a mass of soil resting on an inclined layer of impermeable soil.

EXAMPLE 12-1

Given

1. Figure 12-2 shows a 15-ft cut through two soil strata. The lower is a highly impermeable cohesive soil.

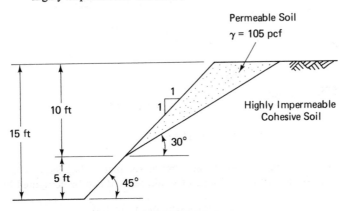

FIGURE 12-2

2. The shearing strength data between the two strata are as follows:

    Cohesion = 150 psf

    Angle of internal friction = 25°

    The mass unit weight of the upper layer = 105 pcf

3. Neglect the effects of soil water between the two strata.

Required

Determine the factor of safety against sliding.

Solution

(see Fig. 12-3)

$$\tan 30° = \frac{10}{A} \qquad A = \frac{10}{\tan 30°} = 17.32 \text{ ft}$$

$$B + 15 = 5 + A \qquad B = 5 + 17.32 - 15 = 7.32 \text{ ft}$$

$$\sin 30° = \frac{10}{L} \qquad L = \frac{10}{\sin 30°} = 20 \text{ ft}$$

The total vertical pressure on the plane *MN*, *W* (i.e., the weight of the triangle of soil *MON*), can be computed by

$$(\tfrac{1}{2})(B)(10)(\gamma) = (\tfrac{1}{2})(7.32)(10)(105) = 3843 \text{ lb/ft}$$

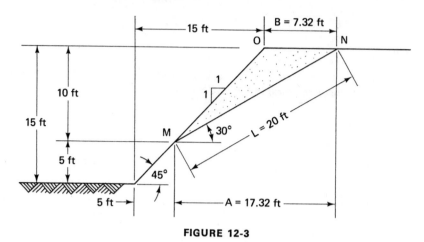

FIGURE 12-3

The sliding force (i.e., shearing stress on the plane MN) is the component of W that is parallel to the plane of weakness and can be computed by

$$(3843)(\sin 30°) = 1922 \text{ lb/ft}$$

Resistance to sliding (i.e., shearing strength on the plane MN) can be computed by

$$cL + W \cos 30° \tan \phi = (150)(20) + (3843)(\cos 30°)(\tan 25°) = 4552 \text{ lb/ft}$$

Factor of safety of the slope against sliding (along the plane of weakness, MN)

$$= \frac{\text{resistance to sliding}}{\text{sliding force}}$$

$$= \frac{4552}{1922} = 2.37 > 1.5 \qquad \text{O.K.}$$

12-3 ANALYSIS OF HOMOGENEOUS SOIL

Two methods are presented in this section for analyzing slope stability for the case of a homogeneous soil. One is known as the *Culmann method* and the other might be called the *stability number method*.

Culmann Method

In the Culmann method, the assumption is made that failure (sliding) will occur along a plane that passes through the toe of the fill [2]. Such a plane is indicated in Fig. 12-4. It can be shown that the safe depth of cut, H,

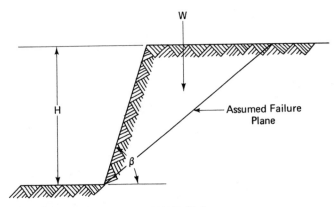

FIGURE 12-4

can be computed for a given soil as a function of the angle β (Fig. 12-4) by the equation [3]

$$H = \frac{4c_d \sin \beta \cos \phi_d}{\gamma[1 - \cos(\beta - \phi_d)]} \tag{12-1}$$

where H = safe depth of cut

c_d = developed cohesion

β = angle from horizontal to cut surface (see Fig. 12-4)

ϕ_d = developed angle of internal friction of the soil

γ = unit weight of the soil

In using Eq. (12-1) to compute safe depth of cut, the developed cohesion (c_d) and the developed angle of internal friction (ϕ_d) may be determined by dividing the cohesion and the tangent of the angle of internal friction by the respective safety factors.

The Culmann method gives reasonably accurate results if the slope is vertical, or nearly vertical (i.e., angle β equal to, or nearly equal to, 90°) [2]. Example 12-2 illustrates the Culmann method.

EXAMPLE 12-2

Given

1. A vertical cut is to be made through a soil mass.

2. The soil to be cut has the following properties:

Unit weight (γ) = 105 pcf

Cohesion (c) = 500 psf

Angle of internal friction (ϕ) = 21°

Required

Determine the safe depth of cut in this soil by the Culmann method, using a factor of safety of 2.

Solution

From Eq. (12-1),

$$H = \frac{4c_d \sin \beta \cos \phi_d}{\gamma[1 - \cos(\beta - \phi_d)]} \tag{12-1}$$

Here,

$$c_d = \frac{c}{F_c} = \frac{500}{2} = 250 \text{ psf} \quad (F_c \text{ is the factor of safety for cohesion})$$

$$\tan \phi_d = \frac{\tan \phi}{F_\phi} = \frac{\tan 21°}{2} = 0.192 \quad (F_\phi \text{ is the factor of safety for } \tan \phi)$$

$$\phi_d = \arctan 0.192 = 10.87°$$

$$\beta = 90° \text{ (vertical cut)}$$

$$H = \frac{(4)(250) \sin 90° \cos 10.87°}{(105)[1 - \cos(90° - 10.87°)]} = 11.5 \text{ ft}$$

Stability Number Method

In the stability number method, a parameter N_s, called the *stability number*, is defined as [4]

$$N_s = \frac{\gamma H}{c} \tag{12-2}$$

where γ = unit weight of soil

H = height of cut (see Fig. 12-5)

c = cohesion of soil

For the embankment illustrated in Fig. 12-5, three types of failure surface are possible. These are shown in Fig. 12-6. In the case of the toe circle, the

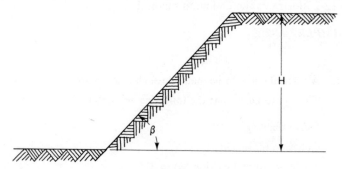

FIGURE 12-5

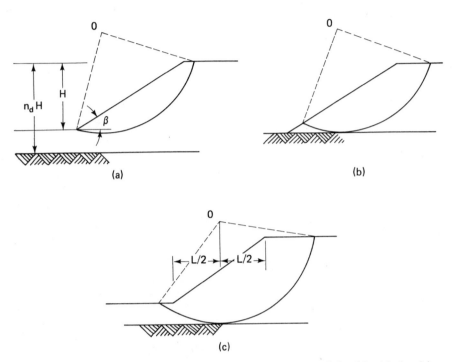

FIGURE 12-6 Types of failure surface. (a) toe circle; (b) slope circle; (c) midpoint cicle. [4][1]

failure surface passes through the toe. In the case of the slope circle, the failure surface intersects the slope above the toe. In the case of the midpoint circle, the center of the failure surface is on a vertical line passing through the midpoint of the slope [4].

Both the type of failure surface and the stability number can be determined for a specific case based on the values of ϕ (angle of internal friction) and β (slope angle, Fig. 12-5). If the value of ϕ is zero, or nearly zero, Fig. 12-7 may be used to determine both the type of failure surface and the stability number. One enters along the abscissa at the value of β and moves upward to the line that indicates the appropriate value of n_d. (n_d is a depth factor and is related to the distance to the underlying layer of stiff material or bedrock and is determined from the relationship indicated in Fig. 12-6a.) The type of line for n_d indicates the type of failure surface, and the value of the stability number is determined by moving to the left and reading from the ordinate. Observation of Fig. 12-7 indicates that if β is greater than $53°$, the failure surface is always a toe circle. Observation also indicates that if n_d is greater than 4, the failure surface is always a midpoint circle [4].

[1] From Tien Hsing Wu, *SOIL MECHANICS.* Copyright © 1976 by Allyn and Bacon, Inc., Boston. Reprinted with permission.

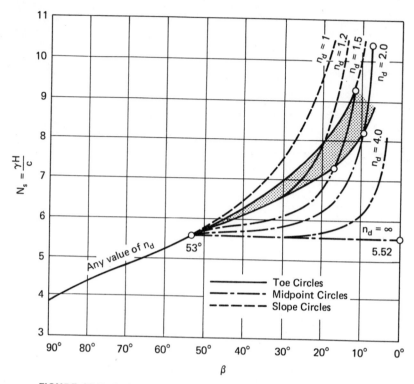

FIGURE 12-7 Stability numbers and types of slope failures for $\phi = 0$. [5, 6]

If the value of ϕ is greater than 3°, the failure surface is always a toe circle [4]. Figure 12-8 may be used to determine the stability number for different values of ϕ [5]. One enters along the abscissa at the value of β and moves upward to the line that indicates the ϕ angle and then to the left where the stability number is read from the ordinate.

Examples 12-3 through 12-5 illustrate the application of the stability number method.

EXAMPLE 12-3

Given

The slope and data shown in Fig. 12-9.

Required

Determine the factor of safety against failure by means of the stability number method.

Solution

Since the given angle of internal friction (ϕ) is 10°, which is greater than 3°, the failure surface will be a toe circle.

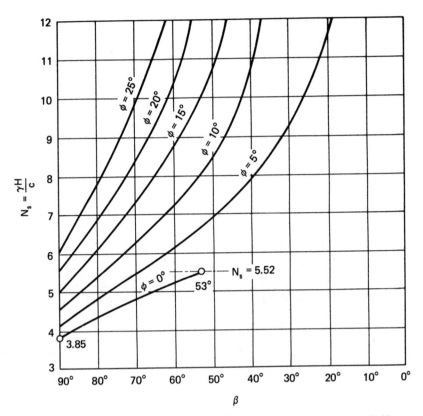

FIGURE 12-8 Stability numbers for soils having cohesion and friction. [5, 6]

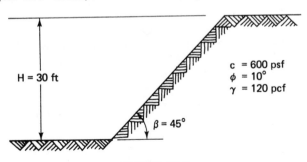

FIGURE 12-9

Try $F_\phi = 1$ (F_ϕ is the factor of safety for tan ϕ):

$$\tan \phi_{\text{required}} = \frac{\tan \phi_{\text{given}}}{F_\phi} = \frac{\tan 10°}{1}$$

$$\phi_{\text{required}} = 10°$$

With $\phi_{\text{required}} = 10°$ and $\beta = 45°$, from Fig. 12-8,

$$N_s = 9.2$$

$$N_s = \frac{\gamma H}{c} \qquad\qquad\qquad\qquad (12\text{-}2)$$

$$\gamma = 120 \text{ pcf}$$

$$H = 30 \text{ ft}$$

$$9.2 = \frac{(120)(30)}{c_{\text{required}}}$$

$$c_{\text{required}} = \frac{(120)(30)}{9.2} = 391 \text{ psf}$$

$$F_c = \frac{c_{\text{given}}}{c_{\text{required}}} = \frac{600}{391} = 1.53 \qquad (F_c \text{ is the factor of safety for cohesion})$$

Since F_ϕ and F_c are not the same value, another value of F_ϕ will be tried.

Try $F_\phi = 1.2$:

$$\tan \phi_{\text{required}} = \frac{\tan \phi_{\text{given}}}{F_\phi} = \frac{\tan 10°}{1.2} = 0.147$$

$$\phi_{\text{required}} = 8.36°$$

With $\phi_{\text{required}} = 8.36°$ and $\beta = 45°$, from Fig. 12-8,

$$N_s = 8.6$$

$$c_{\text{required}} = \frac{(120)(30)}{8.6} = 419 \text{ psf}$$

$$F_c = \frac{c_{\text{given}}}{c_{\text{required}}} = \frac{600}{419} = 1.43$$

Again, F_ϕ and F_c are not the same value; hence, another value of F_ϕ will be tried.

Try $F_\phi = 1.5$:

$$\tan \phi_{\text{required}} = \frac{\tan \phi_{\text{given}}}{F_\phi} = \frac{\tan 10°}{1.5} = 0.118$$

$$\phi_{\text{required}} = 6.73°$$

With $\phi_{\text{required}} = 6.73°$ and $\beta = 45°$, from Fig. 12-8,

$$N_s = 7.9$$

$$c_{\text{required}} = \frac{(120)(30)}{7.9} = 456 \text{ psf}$$

$$F_c = \frac{c_{\text{given}}}{c_{\text{required}}} = \frac{600}{456} = 1.32$$

Again, F_ϕ and F_c are not the same value. Rather than continue a trial-and-error solution, plot the values computed.

$$F_\phi = 1.0 \quad \text{and} \quad F_c = 1.53$$
$$F_\phi = 1.2 \quad \text{and} \quad F_c = 1.43$$
$$F_\phi = 1.5 \quad \text{and} \quad F_c = 1.32$$

As shown in Fig. 12-10, the factor of safety is the point on the curve drawn through the plotted points where F_ϕ is equal to F_c. This point is determined by finding the point where the curve drawn through the plotted points intersects a 45° line as shown in Fig. 12-10. From Fig. 12-10, the factor of safety of the slope against failure is observed to be 1.36.

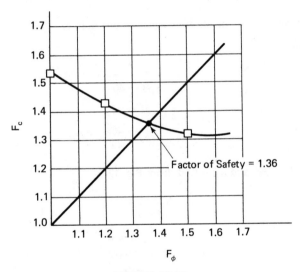

FIGURE 12-10

EXAMPLE 12-4

Given

1. A cut 25 ft deep is to be made in a stratum of highly cohesive soil (see Fig. 12-11).

2. The slope angle β is 30°.

3. Soil exploration indicated that bedrock is located 40 ft below the original ground surface.

4. The soil has a unit weight of 120 pcf, and its cohesion and angle of internal friction are 650 psf and 0°, respectively.

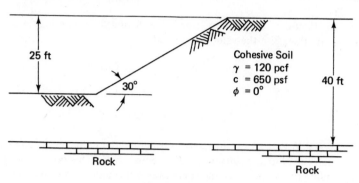

FIGURE 12-11

Required

Determine the factor of safety against slope failure.

Solution

From Fig. 12-6a,

$$n_d H = 40$$

$$H = 25 \text{ ft}$$

$$n_d = \frac{40}{25} = 1.60$$

With $\beta = 30°$ and $n_d = 1.60$, from Fig. 12-7,

$$N_s = 6.0$$

$$N_s = \frac{\gamma H}{c_{required}} \qquad (12\text{-}2)$$

$$\gamma = 120 \text{ pcf}$$

$$H = 25 \text{ ft}$$

$$6.0 = \frac{(120)(25)}{c_{required}}$$

$$c_{required} = \frac{(120)(25)}{6.0} = 500 \text{ psf}$$

$$\text{Factor of safety against failure} = \frac{c_{given}}{c_{required}} = \frac{650}{500} = 1.30$$

EXAMPLE 12-5

Given

1. A cut 30 ft deep is to be made in a deposit of highly cohesive soil that is 60 ft thick and is underlain by rock (see Fig. 12-12).

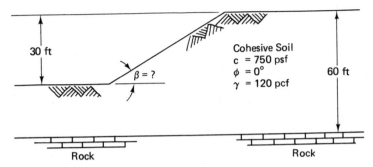

FIGURE 12-12

2. The properties of the soil to be cut are as follows:

$$c = 750 \text{ psf}$$
$$\phi = 0°$$
$$\gamma = 120 \text{ pcf}$$

3. The factor of safety of the slope against failure must be 1.25.

Required

Estimate the slope angle (β) at which the cut should be made.

Solution

From Fig. 12-6a,

$$n_d H = 60$$
$$H = 30 \text{ ft}$$
$$n_d = \frac{60}{30} = 2.0$$

From Eq. (12-2),

$$N_s = \frac{\gamma H}{c_{required}} \qquad (12\text{-}2)$$
$$\gamma = 120 \text{ pcf}$$
$$H = 30 \text{ ft}$$
$$c_{required} = \frac{c_{given}}{F.S.} = \frac{750}{1.25} = 600 \text{ psf}$$
$$N_s = \frac{(120)(30)}{600} = 6.0$$

From Fig. 12-7, with $N_s = 6.0$ and $n_d = 2.0$,

$$\beta = 23°$$

12-4 METHOD OF SLICES

In Sec. 12-3, the assumption was made in the Culmann method that failure (sliding) would occur along a plane that passes through the toe of the slope. It is probably more likely, and observations suggest, that failure will occur along a curved surface (rather than a plane) within the soil. The method of slices, which was developed in the 1920s by Swedish engineers, permits the analysis of slope stability assuming that failure occurs along a curved surface.

The first step in applying the method of slices is to draw to scale a cross section of the slope such as that shown in Fig. 12-13. A trial curved surface

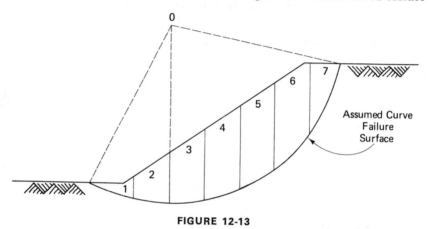

FIGURE 12-13

along which sliding failure is assumed to take place is then drawn. This trial surface is normally approximately circular. The soil contained between the trial surface and the slope is then divided into a number of vertical slices of equal width. The weight of soil within each slice can be calculated by multiplying the volume of the slice by the unit weight of the soil. (This problem is, of course, three-dimensional; however, by assuming a unit thickness throughout the computations, the problem can be treated as two-dimensional.)

Figure 12-14 shows a sketch of a single slice. The weight of the soil within the slice is, of course, a vertically downward force (W in Fig. 12-14). This force can be resolved into two force components—one normal to the base of the slice (W_n) and one parallel to the base of the slice (W_p). It is the parallel component that tends to cause sliding failure. Resistance to sliding is afforded by the cohesion and the friction of the soil. The cohesion force is equal to the cohesion of the soil multiplied by the length of the curved base of the slice. The friction force is equal to the component of W normal to the base (W_n) multiplied by the friction coefficient (tan ϕ, where ϕ is the angle of internal friction).

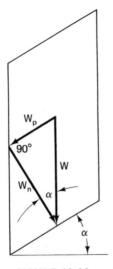

FIGURE 12-14

Since W_p, the component tending to cause sliding of the slice, is equal to W multiplied by sin α (see Fig. 12-14), the *total* force tending to cause sliding of the entire soil mass is the summation of the products of the weight of each slice times the respective value of sin α, or $\sum W \sin \alpha$. Since W_n is equal to W multiplied by cos α, the *total* friction force tending to resist sliding of the entire soil mass is the summation of the products of the weight of each slice times the respective value of cos α times tan ϕ, or $\sum W \cos \alpha \tan \phi$. The *total* cohesion force tending to resist sliding of the entire soil mass can be computed simply by multiplying the cohesion of the soil by the (total) length of the trial curved surface, or cL. Based on the foregoing, the factor of safety can be computed by the equation

$$\text{F.S.} = \frac{cL + \sum W \cos \alpha \tan \phi}{\sum W \sin \alpha} \tag{12-3}$$

(As will be related in Example 12-6, the term $W \sin \alpha$ may be negative in certain situations.)

Computing the factor of safety by the method related above gives the factor of safety for the specific assumed failure surface. It is quite possible that the circular surface selected may not be the weakest, or the one along which failure would occur. It is essential, therefore, that several different circular surfaces be analyzed until the designer is satisfied that the worst condition has been considered.

EXAMPLE 12-6

Given

1. The stability of a slope is to be analyzed by the method of slices.

2. On a particular trial curved surface through the soil mass (see Fig. 12-15), the shearing component (i.e., sliding force) and the normal

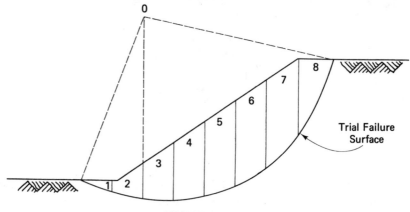

FIGURE 12-15

component (i.e., normal to the base of each slice) of the weight of each slice are as tabulated below.

Slice Number	Shear Component $(W \sin \alpha)$ (lb)	Normal Component $(W \cos \alpha)$ (lb)
1	-63[1]	358
2	-51[1]	1450
3	86	2460
4	722	3060
5	1470	3300
6	1880	3130
7	2200	2270
8	950	91

[1]Since the trial surface curves upward near its lower end, the shearing components of the weights of slices 1 and 2 will act in a direction opposite to those along the remainder of the trial curve, resulting in a negative sign.

3. The length of the trial curved surface is 36 ft.

4. The ϕ angle of the soil is 5° and the cohesion (c) is 400 psf.

Required

Determine the factor of safety of the slope along this particular trial surface.

Solution

From Eq. (12-3)

$$\text{F.S.} = \frac{cL + \sum W \cos \alpha \tan \phi}{\sum W \sin \alpha} \qquad (12\text{-}3)$$

$$c = 400 \text{ psf}$$

$$L = 36 \text{ ft}$$

$$\sum W \cos \alpha = 358 + 1450 + 2460 + 3060 + 3300 + 3130 + 2270 + 91$$

$$= 16,119 \text{ lb}$$

$$\phi = 5°$$

$$\sum W \sin \alpha = -63 - 51 + 86 + 722 + 1470 + 1880 + 2200 + 950$$

$$= 7194 \text{ lb}$$

$$\text{F.S.} = \frac{(400)(36) + (16,119) \tan 5°}{7194} = 2.20$$

12-5 PROBLEMS

12-1 Figure 12-16 shows a 20-ft cut through two soil strata. The lower is a highly impermeable cohesive clay. The shear strength data between the two strata are as follows:

$$c = 220 \text{ psf}$$

$$\phi = 12°$$

The unit weight of the upper layer is 110 pcf. Determine if a slide is likely by computing the factor of safety against sliding. Neglect the effects of soil water.

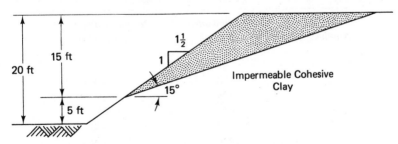

FIGURE 12-16

12-2 A vertical cut is to be made in a deposit of homogeneous soil mass. The soil mass to be cut has the following properties: The unit weight of the soil is 120 pcf, cohesion (*c*) is 350 psf, and the angle of internal friction (ϕ) is 10°. It has been specified that the factor of safety against sliding must be 1.50. Using Culmann's method, determine the safe depth of cut.

12-3   Determine the factor of safety against a slope failure by means of the stability number method for the slope shown in Fig. 12-17.

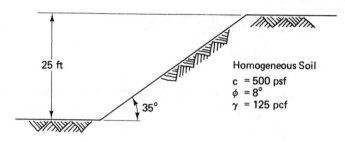

25 ft

35°

Homogeneous Soil

c = 500 psf
ϕ = 8°
γ = 125 pcf

FIGURE 12-17

12-4   A cut 20 ft deep is to be made in a stratum of highly cohesive soil that is 80 ft thick and is underlain by bedrock. The slope of the cut is 2:1 (i.e., 2 horizontal to 1 vertical). The unit weight of the clay is 110 pcf, and its c and ϕ values are 500 psf and 0°, respectively. Determine the factor of safety against slope failure.

12-5   A cut 25 ft deep is to be made in a deposit of cohesive soil with $c = 700$ psf, $\phi = 0°$, and $\gamma = 115$ pcf. The soil is 30 ft thick and is underlain by rock. The factor of safety of the slope against failure must be 1.50. At what slope angle (β) should the cut be made?

12-6   The stability of a slope is to be analyzed by the method of slices. On a particular trial curved surface through the soil mass, the shearing and the normal components of the weight of each slice are as tabulated below. The length of the trial curved surface is 40 ft. The cohesion c and ϕ angle of the soil are 225 psf and 15°, respectively. Determine the factor of safety of the slope along this trial surface.

Slice Number	Shear Component ($W \sin \alpha$) (lb)	Normal Component ($W \cos \alpha$) (lb)
1	−38	306
2	−74	1410
3	124	2380
4	429	3050
5	934	3480
6	1570	3540
7	2000	3210
8	2040	2190
9	766	600

References

[1] GEORGE F. SOWERS, *Introductory Soil Mechanics and Foundations: Geotechnical Engineering*, 4th ed., Macmillan Publishing Co., Inc., New York, 1979.

[2] DONALD W. TAYLOR, *Fundamentals of Soil Mechanics*, John Wiley & Sons, Inc., New York, 1948. Copyright © 1948, by John Wiley & Sons, Inc. Reprinted by permission of John Wiley & Sons, Inc.

[3] MERLIN G. SPANGLER AND RICHARD L. HANDY, *Soil Engineering*, 3rd ed., Intext Educational Publishers, New York, 1973. Copyright 1951 © 1960, 1973 by Harper & Row, Publishers, Inc. Reprinted by permission of the publisher.

[4] TIEN HSING WU, *Soil Mechanics*, Allyn and Bacon, Inc., Boston, 1976. Copyright © 1976 by Allyn and Bacon, Inc., Boston. Reprinted with permission.

[5] KARL TERZAGHI AND RALPH B. PECK, *Soil Mechanics in Engineering Practice*, John Wiley & Sons, Inc., New York, 1967. Copyright © 1967, by John Wiley & Sons, Inc. Reprinted by permission of John Wiley & Sons, Inc.

[6] D. W. TAYLOR, "Stability of Earth Slopes," *J. Boston Soc. Civil Eng.*, **24** (1937).

Index